ANTI-SLAVERY *and the* UNDERGROUND RAILROAD *in* FAIRFIELD, IOWA

RORY GOFF

Milton LeRoy Goff IV © 2018

Merrymeeting Archives LLC
PO BOX 1058
Fairfield, IA   52556

Ordering Information:
Special discounts are available on quantity purchases by corporations, associations, educators, and others. For details, contact the publisher at the above listed address.

U.S. trade bookstores and wholesalers:
Please contact Merrymeeting Archives LLC

*"...it is necessary to KNOW what has to be done,
to WILL what is required, to DARE what must be attempted
and to KEEP SILENT with discernment."* -- *Eliphas Levi*

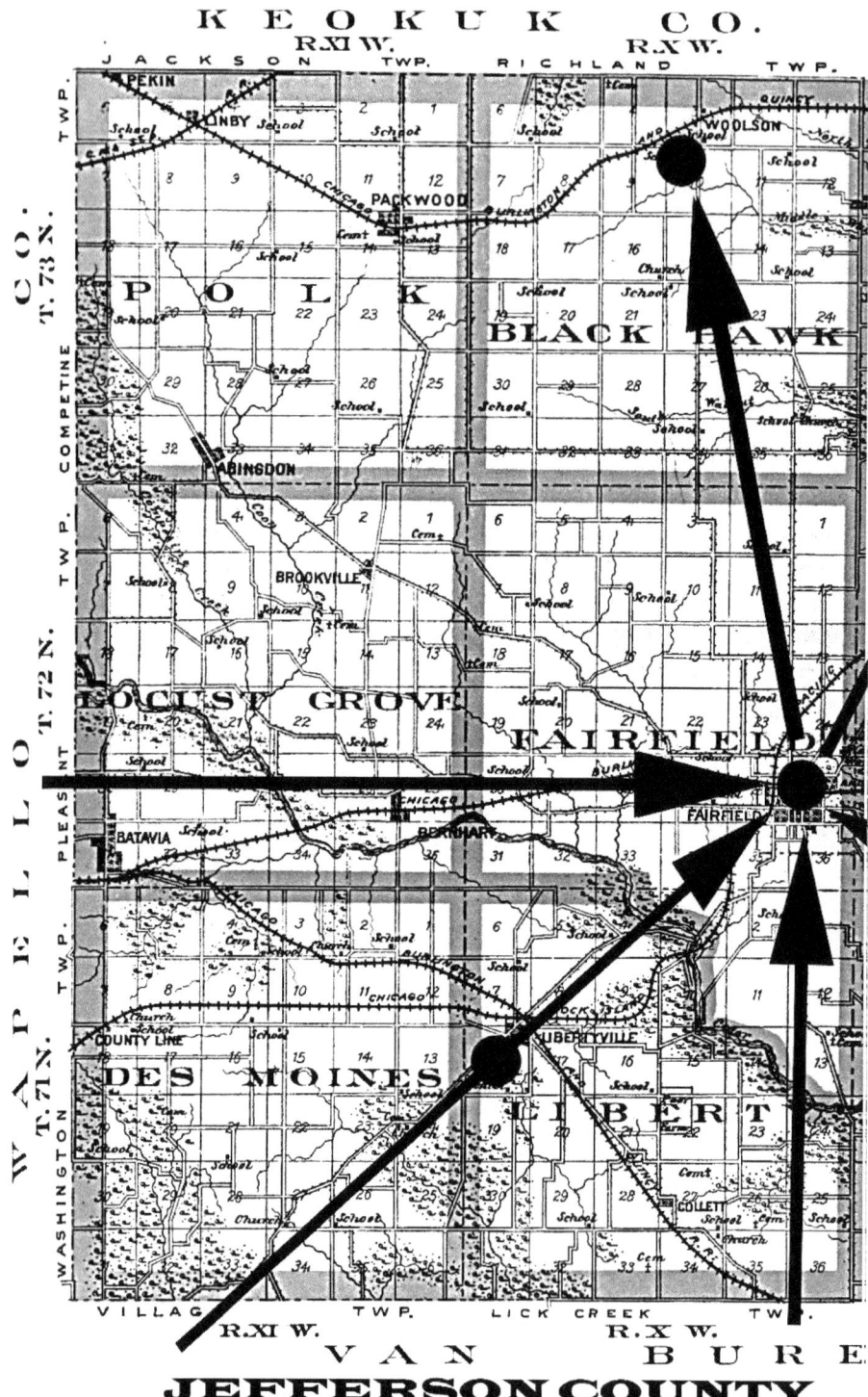

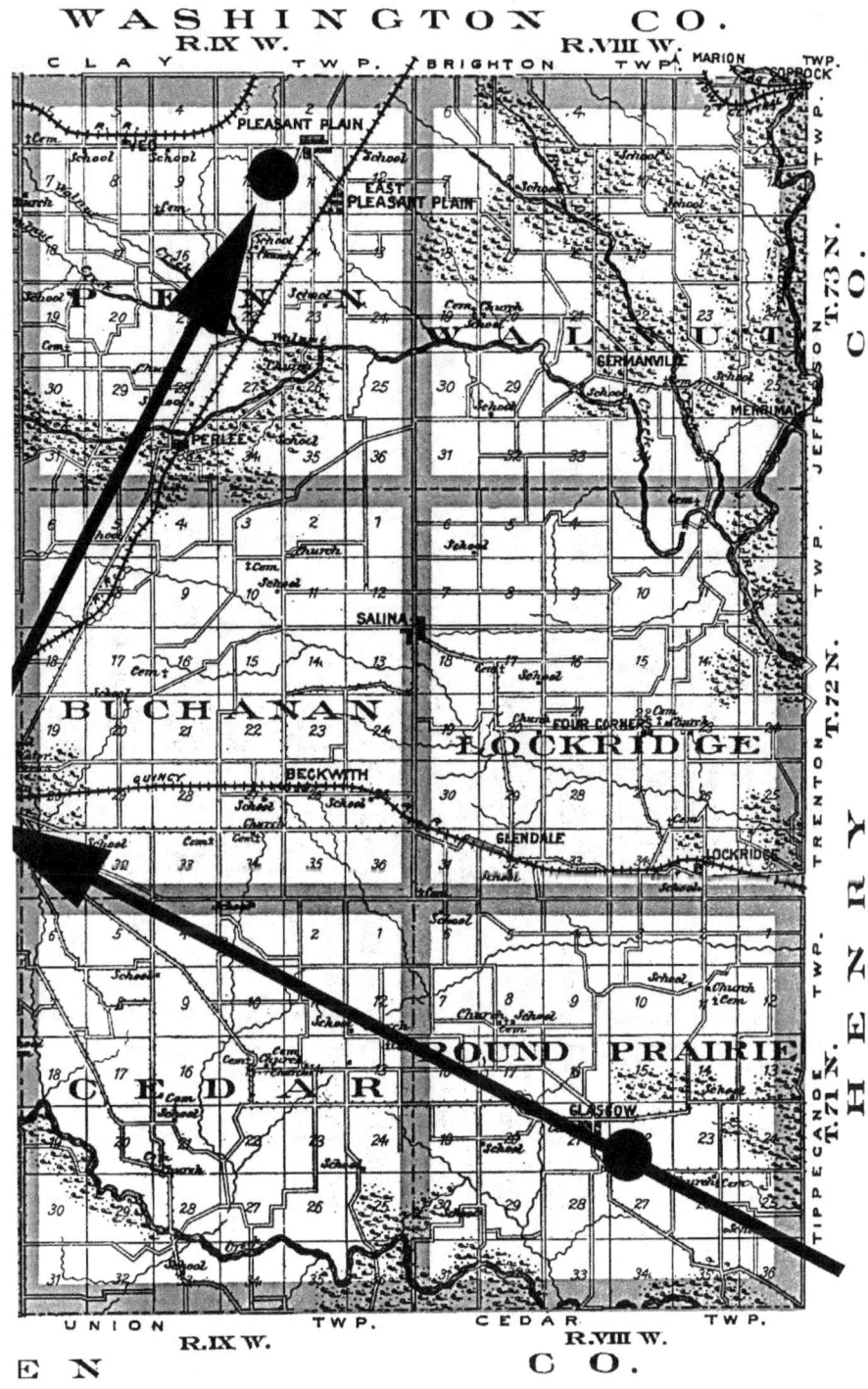

TOWNSHIP MAP (1904)

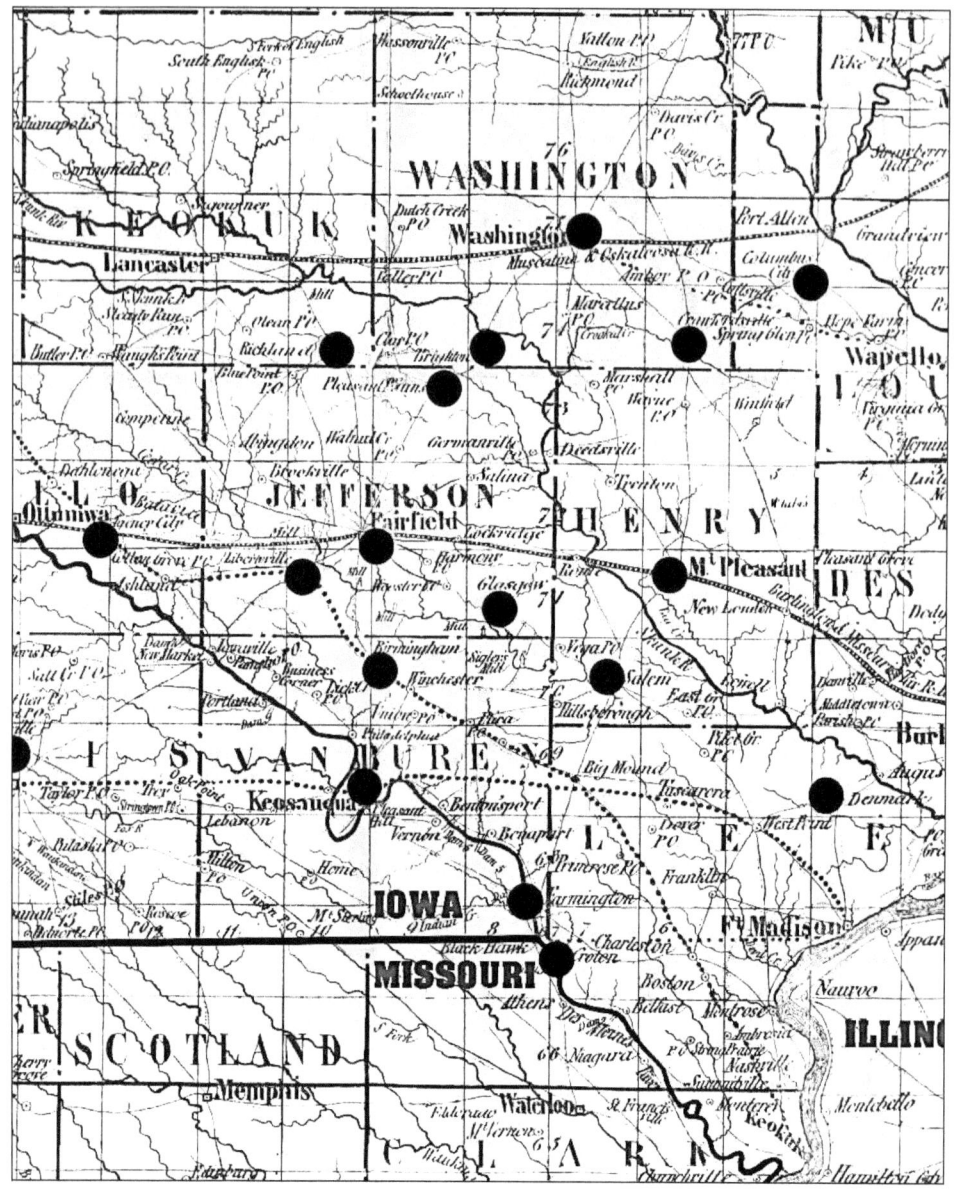

Previous page: Prominent Underground Railroad Routes through Fairfield, Jefferson County, Iowa.

This page: Some prominent UGRR towns in Southeast Iowa, west to east: Drakesville, Ottumwa, Libertyville, Richland, Keosauqua, Birmingham, Fairfield, Pleasant Plain, Brighton, Glasgow, Farmington, Croton, Washington, Salem, Mt. Pleasant, Crawfordsville, Columbus, and Denmark. Fairfield was about 30 miles north of the slave state of Missouri.

Introduction

This book outlines the tumultuous history of the growth of the anti-slavery movement in Fairfield and Jefferson County, Iowa. From a tiny, despised minority at Fairfield's founding in 1839, the movement grew into a popular progressivism which encouraged and supported President Abraham Lincoln's Emancipation Proclamation outlawing slavery in the Southern states on January 1, 1863, and declaring that all slaves within the rebellious states "are, and henceforward shall be free."

Fairfield, Iowa is home to hundreds who have stood up for freedom and human rights, defying a tyranny which for centuries had crushed the bodies, souls, and spirit of people of darker skin: an oppression mostly condoned by popular opinion, government, and many churches. The companion volume, *Who's Who in the Anti-Slavery and Underground Railroad Networks in Fairfield, Iowa,* gives biographies of each of those heroes, and when possible shows each person's connections to other anti-slavery luminaries, both locally and nationally.

In contrast, this volume examines how and why Fairfield's anti-slavery heroes and heroines came to take the stand for freedom that they did. Probably as soon as Fairfield was founded, and more certainly within its first eight years, some Jefferson County pioneers were transcending their own immediate needs of frontier survival, and taking great risks to secretly and illegally feed, clothe, shelter, and transport fugitive slaves: strangers whom they had never met before and would never see again. Many of these heroes and heroines were deeply spiritual; doubtless some felt they were following Jesus, who had said, "Inasmuch as ye have done it unto one of the least of these my brethren, ye have done it unto me," or St. Paul, who wrote, "Be not forgetful to entertain strangers: for thereby some have entertained angels unawares."[1]

This work sprang out of a decade's research into reports that Ulysses S. Grant wintered in Fairfield during the unofficial civil war over slavery in "Bleeding Kansas." Investigation gradually uncovered a large cast of characters with a dense network of antislavery and Underground Railroad (UGRR) connections in Jefferson County, many with surprising ties to U. S. Grant himself, as well as to people and places important to him in Pennsylvania and Ohio. The work on Grant in Fairfield itself sprang from yet another project: investigating the prop-

---

[1] KJV, Matthew 25:40, Hebrews 13:12.

erties around Fairfield's central square from 1839 to now. That led into early genealogies, deeds, letters, censuses, directories, and old Fairfield newspapers: which in 1882 first broke the story of U. S. Grant's residence in Fairfield, Iowa, and sent us on that decade-long "detour."

That original research revealed that Fairfield, Iowa was in all its diversity a microcosm reflecting the nation as a whole in its changing attitudes toward slavery, and was strategically located to play a significant role in both the UGRR from Missouri northeast to Canada, and the Free-Soil immigration west into "Bleeding Kansas." We hope that our analysis of the growth of Fairfield's anti-slavery movement from a tiny and scorned minority into a popular progressivism that overthrew a deeply entrenched oligarchy fattening on human suffering, may inspire us all to heartily support freedom and human rights, and to firmly oppose any oppression still present today.

As George Santayana famously said, "Those who cannot remember the past are condemned to repeat it." We thank the countless heroes over the past 200 years, both known and unknown, who have remembered, and who have carefully preserved and transmitted the fragile and fleeting memories of what *was*, in order to help us deal with what *is*.

Among innumerable others, we wish to thank Director Rebecca Huggins of the Fairfield Public Library and all of its librarians; local historians Lee Gobble, Verda Baird, Douglas Hamilton, Richard K. Thompson, Gene Luedtke, Mark Shafer, and Herb Shafer; Kelly Spees, Liz Hickenbottom, Christ Carter, Carol Miner, Charlotte Fleig, and Sandy Hoskins of the Jefferson County Recorder's office; the scholars and librarians of the State Historical Society Archives in Iowa City and Des Moines, and the creators, maintainers, and scholars of many websites: especially Ancestry.com for detailed genealogical records and censuses, Findagrave.com for extensive burial and obituary data, County Coordinator Joey Stark of Iagenweb.org for the 1871 Jefferson County Township maps and cemetery records, Newspaperarchive.com for an impressive online collection of early periodicals, and eBay for facilitating acquisition of old letters, photographs, and memorabilia.

We wish to thank Richard DeAngelis, Lawrence Eyre, and everyone at Fairfield Productions for their heartfelt passion and expertise, exposing us to new resources and experts. We are immensely grateful to UGRR scholars Lowell Soike, Owen W. Muelder, and David Holmgren for their invaluable assistance, suggestions, and corrections. All errors still remaining are, of course, ours alone; we welcome any corrections.

A Fairfield Boy on the Underground Railroad

You don't feel like a hero. You're only a kid, and your heart is racing as you heave the saddle onto your horse's blanketed back, tighten the girth, and hang your little old muzzle-loading shotgun by a strap from the pommel. The flintlock is primed and loaded, and you hang it breech-down. You swing up into the saddle and leave the farm, trotting south back down the road in the deepening dusk to the bottom of the hill, almost to the cemetery. There, with no one in sight, you cut left into the cool woods by Crow Creek.

You reach the old "haunted" slaughterhouse, dilapidated and deserted, where that afternoon you had slipped in on foot and fed the newest fugitive. You have seen runaways who were cruelly beaten before, but somehow they always seem optimistic. As always, this one somehow managed to entirely devour the whole loaf of home-made bread, and the large chunk of cooked meat, and all six or eight boiled potatoes from the half-filled grain-sack your mother gave you in the kitchen back home. Others had been present, so she had told you to go salt the stock out on the farm.

You knew what she meant, and you had taken the sack and walked north out of town past the cemetery, making sure you weren't followed, and whistling to keep up your courage. By helping runaway slaves, your family is "Negro-stealing." The slaves come to Fairfield by way of Salem, from Missouri, you think, but your parents do most of the thinking; you just feed the fugitives and move them along. If your family is caught, they will get huge fines and years in prison, maybe even die in chains. You will lose everything. Plenty of Southern-sympathizing neighbors would happily turn you in to the slave-catchers and U. S. marshals to claim the reward. You are not yet twelve, but you believe in your family's cause; you believe everyone deserves freedom, and you are proud to help.

The runaway comes out of the slaughterhouse and climbs onto your horse just behind the saddle. He tends to clasp you too tightly around your waist, so you whisper, "Don't hold on so, Jim!" as you pick your way east and north through the woods, cautious and quiet.

You emerge from the woods onto the road in the dark of night but there is a moon, and as instructed, you lean over your horse's neck and hug the tree-line as you trot a little faster. "Jim" starts to laugh aloud with the giddy anticipation of freedom, and you hiss, "Shut up!" and elbow him into a brooding silence. A twelve-mile ride northeast will get you to the Friends in Pleasant Plain, who will take your passenger on his next leg to Canada and freedom. Easy enough, if you don't encounter any slave-catchers.

But this time, you do.

You have just nudged your passenger to relax his clasp yet again as you round a sharp corner and come upon two horsemen talking in the road. You rein in hard, setting your horse back on his haunches and unseating "Jim," who disappears behind you. Your shotgun hits the roadbed, unlimbers from the saddle horn, falls forward and goes off with a deafening BOOM! as a stream of fire lights the sky. Panicked, your horse bounds forward and races flat out, ears back, and your hat flies off in the wind as you cling tightly for dear life.

You finally calm your horse a mile or more down the road. Your passenger is long gone, finding his own way north, you hope. You swallow hard, tasting the bitterness of failure. You ride back to the farm, stable your horse, and walk home at daybreak. You remember to buy some steak at the butcher's first, as your parents had instructed, and you show it to the spy watching your house when he demands to know why you are out so early.

Later that morning on your way to school, you see two slave-catchers telling six or eight townsfolk near the livery barn how a "dozen or more men" had come suddenly upon them, firing their guns and badly crippling one of their "hosses," and then escaped down the road before the slavers could recover from their surprise. As evidence, they are holding your little old shotgun.

Your family members are gunsmiths, and you are acquainted with every gun in the neighborhood. The locals call you over, asking if you know whose gun it is. You look your shotgun over carefully, stalling for time, and then name a pro-slavery neighbor. They scoff and let you go.

That afternoon a venerable old Quaker from Pleasant Plain drives his buggy slowly up to your house, calls you out, looks around to make sure you're alone, and then hands you your battered old hat, saying, "Son, is this thy cap?"

"Yes, Uncle Josiah."

"Well, thee should be more careful where thee hangs it up," he says, and drives away.

Because you failed, this was your last run on the Underground Railroad. But you will keep your family's secret well. You do not speak *of it for more than 63 years, until you are 74, and everyone else involved is long dead.*

You are Christian S. Byrkit, and you later served ten years as Iowa's deputy secretary of state, but today you are best remembered for your earlier work with your family in the cause of freedom. Your rare first-hand account illumines a few of the unsung heroes of the Underground Railroad in Fairfield, Iowa, and uncovers their connections with the Quakers of Salem, Richland and Pleasant Plain.

Thank you for your service.

* * *

Three Magnitudes of Anti-Slavery

Before the American Civil War, three magnitudes of antislavery ardor pierced the gloom of eternal servitude:

- *Free-Soilers*, who *knew* that slavery must be curbed, and opposed slavery's expansion into the Western Territories,

- *Abolitionists*, the smaller number who *willed* that slavery must be abolished, and favored emancipation of all slaves, either gradually or immediately, with or without compensation to the "owner," and

- *Underground Railroad (UGRR) operators*, the still smaller number who *dared* to take immediate action, and broke federal law and risked their reputations as well as ruinous fines, imprisonment, and even death, to follow a higher law and help a stranger to freedom.

Free-Soilers and abolitionists could and did speak out, but for Underground Railroad operators, *keeping silence* was crucial for the safety of all involved, and some operators were never suspected even of being Free-Soilers, let alone abolitionists. An almost-total absence of written records often forces us to piece together UGRR operations from snippets of circumstantial evidence.

Even among UGRR operators there were different degrees of commitment, from a one-time giver of food or directions, all the way up to veteran "conductors" like New York's Harriet Tubman and Ohio's John P. Parker, angels of blazing courage who repeatedly risked their lives on forays into the hostile South to free dozens or hundreds of the enslaved; or "directors" like Ohio's Reverend John Rankin and Indiana's Levi Coffin, luminaries who cared for thousands of fugitives over decades despite constant harassment, threats, lawsuits, attacks, and bounties placed upon their heads.

The Underground Railroad was not an Underground Railroad

While freedom-seekers sometimes hid underground in basements and occasionally traveled on trains, the Underground Railroad was generally neither underground nor a railroad. Rather, it was a loosely-connected network of safe-houses ("stations" or "depots") and their compassionate homeowners ("station masters"), by which the conductors secretly led fugitives ("passengers" or "cargo") step by step from the slaveholding South into the North's free states. After the 1850 Fugitive Slave law imperiled Blacks even there, they traveled all the way to Canada. UGRR "agents" helped fugitives find their way onto the Railroad; "stockholders" supported the UGRR with money, and "managers" supplied fugitives with food, medicine, clothing, money, and directions, and arranged transportation. Travel was mostly at night, by foot or on horseback, or by day under concealment in a wagon, or even in a buggy while wearing long-sleeved dress, gloves, bonnet, and veil.

Routes to Freedom through Southeast Iowa

In Southeast Iowa, the Congregationalist town of Denmark in Lee County and the Quaker settlement of Salem in Henry County were the two oldest and best-known Underground Railroad towns, but Fairfield soon became a significant hub of UGRR activity, with routes leading through it from the slave state of Missouri via the Iowa towns of Denmark, Salem, and Glasgow in the southeast, Keosauqua and Winchester or Birmingham in the south, Drakesville and Libertyville in the southwest, and Ottumwa in the west.

Freedom-seekers reportedly would come north to Fairfield along creek-beds, from Turkey Run (originally "Nigger Run") near Birmingham, thence along Big Cedar Creek to Crow Creek, where the "haunted" slaughterhouse the Byrkit family used as a UGRR station once stood. From Fairfield, fugitives often went north and east to Richland, Clay, or Pleasant Plain, and thence northeast via Brighton, Washington, Crawfordsville, and Columbus, to cross the Mississippi River into Illinois at Muscatine or Davenport.[2]

East of Fairfield, a major route from Athens, Missouri led through Iowa via Farmington or Croton north to Salem, and from there either north to Mount Pleasant and Crawfordsville, or northeast to Pleasant Green and Mediapolis, or east to the Mississippi at Burlington. Another led via Farmington or Croton east to Denmark and Burlington.

West of Fairfield, one could go north from Downing or Lancaster, Missouri through Iowa via Bloomfield to Drakesville and either north to Ottumwa or northwest to Blakesburg, then either north to Eddyville, or north-northeast to Kirkville and Fremont, or northeast to Hedrick, and then east through Richland.

The general flow of UGRR traffic was north and east through Iowa, but the routes were fluid. They did not always lead north through Fairfield from Birmingham and Libertyville. One conductor reportedly led freedom-seekers from the Fordyce tavern—just southeast of Birmingham near Winchester and Stockport—along a route slightly south of east to Salem; another brought them from Cincinnati in Appanoose County northeast through Drakesville to Libertyville, and from there east-southeast to Salem. Young UGRR conductor Christian S. Byrkit reported that his parents, who were Quaker and Presbyterian turned Methodist, sheltered refugees coming northwest by west from the Quakers' Salem to Fairfield, and thence north or north-northeast to the Quaker settlements of Richland or Pleasant Plain. Using station masters and conductors they could thoroughly trust, operators shifted fugitives between safe-houses like a human shell game to evade the federal marshals and slave-catchers.

Some UGRR Operators around Fairfield

Perhaps the Byrkits worked with alleged UGRR operator Daniel Mendenhall, who was another Quaker-turned-Methodist. He would

---

[2] For the route from Keosauqua through Winchester or Birmingham to Fairfield, see W. P. A. Federal Writers Project, American Guide Series: Van Buren County, Iowa, 1940, pp. 23-24. "Crow Creek" was named perhaps for its many crows, or perhaps for "Jim Crow," Chris Byrkit's term for all his Black passengers by the creek.

teach the Byrkit boys gunsmithing, and had prominent relatives in Salem. From 1856 he would have other dedicated UGRR relatives in Fairfield: the Quakers Rachel (Coppock) and Benjamin Pierce, who also reportedly conducted freedom-seekers to Pleasant Plain. And perhaps the Byrkits worked with alleged UGRR operator Joseph C. Cooper, a fervent member of UGRR manager Rev. Asa Turner's church in Denmark where Joseph's uncle William B. Cooper was also a UGRR operator. After being ordained in 1853, Rev. Joseph established Congregational Churches both in Salem and in Glasgow, midway between Salem and Fairfield, and next year worked with Fairfield's abolitionist Rev. Levin B. Dennis, the Byrkits' minister. We may never know for certain who the Byrkits' UGRR partners were. Even when Christian finally spoke out 63 years later, beyond his immediate family he still shielded the identity of the other UGRR operators in his old network; he kept silence to the end. Secrecy was the life-blood of the UGRR.[3]

Despite the pervasive secrecy, we do know of a number of additional UGRR operators around Fairfield. Among them, John H. B. Armstrong worked steadily on Iowa's UGRR for decades, first from about 1838 southeast of Fairfield in Croton, Lee County, serving the Denmark and Salem UGRR routes with his first wife, Sidney (Henkle). In 1852 he moved southwest of Fairfield to Cincinnati, Appanoose County, where with Calverts, Fulchers, Holbrooks, McDonalds, Stantons, and others he served on the UGRR with his second wife, Isabella (Shepherd) Frush. Her nephew George Frush married John and Sidney's daughter Mary Jane (Armstrong) in 1856, and lived just northeast of Fairfield in Buchanan Township, probably on the Pleasant Plain Road. Perhaps the Frushes sometimes brought John's "passengers" on their journey's next leg, through Fairfield and on to Pleasant Plain.[4]

East of Fairfield was the UGRR's Congregational patriarch Rev. Asa Turner of Denmark, whose church included operators William and Joseph Cooper as well as the Sackett, Shedd, and Trowbridge families. Asa Turner would inspire UGRR operators in Fairfield and westward

---

[3] The State Historical Society of Iowa's Iowa Freedom Trail Grant Project lists Daniel Mendenhall and Joseph and William Cooper as UGRR operators, but as yet without documentation. For William Cooper as UGRR operator, see William Henry Perrin, ed., "William Brewster Cooper," *History of Effingham County, Illinois*, Chicago: O. L. Baskin & Co., 1883, part 2, p. 11.

[4] For J. H. B. Armstrong as UGRR operator in Fayette Co., Ohio, and in Lee and Appanoose Counties, Iowa, see Wilbur H. Siebert, *The Underground Railroad from Slavery to Freedom*, New York: The MacMillan Co., 1898, pp. 42-43.

throughout Iowa. Ten miles west of Denmark were the UGRR Quakers of New Garden, including the Kellums, Pickards, Bonds, and Hoags. On the UGRR in and around the Quaker town of Salem nearby were the Fraziers, Garretsons, Duvall W. Henderson, and Henderson Llewelling. In Mt. Pleasant's UGRR were the Howes and the Dugdales.

West of Fairfield was UGRR station master Samuel S. Norris, Ottumwa's first Congregational deacon, whose son-in-law Rev. Benjamin A. Spaulding would establish anti-slavery churches at Agency, Oskaloosa, Eddyville, and Ottumwa, often in concert with Rev. Julius A. Reed of Fairfield, and with Rev. George B. Hitchcock. George was Rev. Reed's deacon at Fairfield, then was pastor at Oskaloosa and Eddyville, and finally was a prominent UGRR station master near Lewis in Cass County.[5]

South of Fairfield, Benjamin F. Pearson maintained a UGRR station in Keosauqua; he was a cousin of pioneering Quaker UGRR operator Benjamin Lundy, and also of Fairfield's Martha Pike, Mrs. Robert F. Ratcliff. Messrs. Holbert and Forbes reportedly served the UGRR in Keosauqua: James William Holbert lived in Keosauqua in 1852, but was in Des Moines Township, Jefferson County in 1856-60, and Dr. Alfred Forbes lived in Farmington Township; his daughter Almeda married Henry Henkle, UGRR station master Sidney (Henkle) Armstrong's first cousin.[6]

Coming north from Keosauqua toward Fairfield, Newton and Esther Calhoun were UGRR station masters in Winchester, where the Fordyce Tavern was another reputed station, and UGRR operators in

---

[5] For S. S. Norris on the Ottumwa UGRR, see *Portrait and Biographical Album of Wapello County, Iowa...*, Chicago: Chapman Bros., 1887, pp. 464-465, and *The Advance*, vol. 45, no. 1945 (Feb. 19, 1903). For Benjamin Spaulding's work in organizing the churches at Agency, Ottumwa, Oskaloosa, and Eddyville, see *The Congregational Quarterly*, vol. 10, no. 2 (April 1868), pp. 215-216. For Julius Reed in Oskaloosa and assisting Rev. Spaulding in organizing the church at Eddyville, see *The Home Missionary*, vol. 18, no. 1 (May 1845), p. 5. George B. Hitchcock was Rev. Reed's deacon at Fairfield (1844) and then pastor at Oskaloosa (1844-48) and Eddyville (1847-53) before operating a UGRR station near Lewis, Cass Co., Iowa.

[6] In Keosauqua, Messrs. Holbert and Forbes "helped many to places of refuge, and supplied them with food and means of travel whenever possible." Three houses in Keosauqua were considered UGRR stations: the "old Pearson House, the old Holden Home, and the old Otto White House." Fugitives were "often taken from Keosauqua to Winchester, then to another station in Birmingham, then to Fairfield." Compiled by Workers of the Writer's Program of the W.P.A. in Iowa, Thomas L. Keith, publisher, Farmington, Iowa, 1940, pp. 23-24.

Birmingham were Presbyterian Reverend David Lindsay, Samuel and Ebenezer Gould, William Collier, Peter Groesbeck, and Hiram B. Barnes. Jacob T. and Selina Lamb kept a UGRR station near Libertyville, and Selina's cousin Erastus Nulton and his half-brother Arthur Corner brought freedom-seekers to the Lambs from their UGRR station near Drakesville, midway between Cincinnati and Libertyville.[7]

Going north from Fairfield, Allen Stalker was a UGRR manager between Fairfield and alternately Richland or Pleasant Plain. Allen's second cousin Jesse Hinshaw reportedly maintained a UGRR station in his house in Woolson, about 12 miles north of Fairfield near Richland. UGRR operators in Pleasant Plain included Dr. Thomas S. Mealey and Charles Osborn Stanton. Manning Mills, miller Samuel B. McKain, Alfred Meacham, and Henry Morgan conducted fugitives from Pleasant Plain, Clay, or Richland on to Washington, where they would meet the UGRR brothers John and Martin C. Kilgore, or Andrew Kendall. In Crawfordsville, freedom-seekers could rest at station master Samuel Rankin's hotel, known as the "House of All Nations."[8]

---

[7] "Vurnum Saunders Calhoun," *Portrait and Biographical Album of Jefferson and Van Buren Counties, Iowa* ..., Chicago: Lake City Publishing Co., 1890, p .442; "Newton Calhoun," *The History of Van Buren County, Iowa...*, Chicago: Western Historical Co., 1878, p. 593. For Messrs. Gould, Collier, Groesbeck, and Lindsay on Birmingham's UGRR, see Wilbur Henry Siebert, *The Underground Railroad in Iowa* [1890-1940], 2 vols., Allred-Boulton Family Album, BL 379 f.1, State Historical Society of Iowa, Iowa City: For Erastus Nulton, Arthur Corner, and Selina (Byers) Lamb, see Wilbur Siebert, *The Underground Railroad in Iowa*, Ohio Historical Society Collections, MSS 116Av, 1934, p. 24.

[8] For UGRR operators in Jefferson Co., see State Historical Society of Iowa's Iowa Freedom Trail Grant Project; Rory Goff, *Who's Who in the Anti-Slavery and Underground Railroad Networks in Fairfield, Iowa*, Fairfield, Iowa: Merrymeeting Archives, 2018. In Washington, John and Martin C. Kilgore and Andrew Kendall were UGRR operators: For the Kilgore brothers' taking fugitives to Col. Samuel Rankin in Crawfordsville, see O. A. Garretson, "Traveling the Underground Railroad in Iowa," *Iowa Journal of History and Politics*, Vol. 22 (1924), pp. 451-452. For Andrew Kendall as station master in Washington, Iowa, see Alex R. Miller, "Helping the Fugitive Slave Thru Iowa: Old Kendall Home in Washington, Built of Brick Carried from Burlington[,] was Station on Famous 'Underground Railway' Just Before the Civil War," *The Burlington Hawk-Eye*, May 15, 1921, p. 15 (part 2, p. 1), cols. 1-7.

Abolitionism and the Slavery Addiction

Reverend Henry E. Wing, who healed the Fairfield, Iowa Methodists' split over slavery in 1876, had a harshly abolitionist father and a gentle, Southern-sympathizing mother who felt that Blacks were better off enslaved. Henry took after his mother and before the Civil War he was sure that if the occasion arose, he would return a freedom-seeker to slavery. However, while teaching at a seminary in Vermont, he learned a young Black friend of his was a fugitive now closely pursued by his old "master," and Henry's "quick wit and savings" paid his friend's way to Canada. "But his action disturbed him. He was a lawbreaker, and to his consternation he knew he would do the same thing again."[9]

As in Henry's case, a sudden upsurge of compassion might move a pro-slavery citizen or even a slave-catcher to break the law and help a distressed fugitive escape, but in general, UGRR operators were at least secretly both Free-Soil and abolitionist, while a Free-Soiler might not be an abolitionist, let alone a UGRR operator. Up until the mid-1850s, calling even a moderate Northerner "abolitionist" was an insult.[10]

A quarter-century earlier, many Southerners had still admitted slavery was an evil to be abolished eventually. But the phenomenal profitability of slavery coupled with the discovery in 1831 that Northerners were actually helping slaves escape along a secret "Underground Road" or "Underground Railroad," followed by the British Empire's abolition of slavery in 1833, and by Theodore D. Weld's abolitionist agitation in Ohio in 1834, resulted both in secret Southern Rights Clubs to protect and expand slavery, and in John C. Calhoun's claims first, that the federal government had no authority over individual states—with the corollary that they had the right to secede—and second, that Northern cap-

---

[9] Ida M. Tarbell, *A Reporter for Lincoln: Story of Henry E. Wing, Soldier and Newspaperman*, N.Y.: The Book League of America, 1929, pp. 33-36.

[10] Seasoned slave-catcher Chancey Shaw apprehended the exhausted Eliza in Ripley, Ohio in 1838 as she arose from the semi-frozen Ohio River with her babe in her arms—but, moved by her heroism and despair, said, "Any woman who crossed that river carrying her baby has won her freedom," and ushered her to Rev. John Rankin's house and safety: Ann Hagedorn, *Beyond the River: The Untold Story of the Heroes of the Underground Railroad*, New York: Simon & Schuster, 2004, pp. 135-138. One youthful slave-catcher actually became a UGRR station master: the Virginian James C. Jordan of West Des Moines, Iowa. See James Connor, "The Antislavery Movement in Iowa," *Annals of Iowa*, vol. 40, no. 5 (Summer 1970), p. 352, http://ir.uiowa.edu/annals-of-iowa/vol40/iss5/5/.

italism was as much a form of bondage as Southern slavery, and only time would tell which system was superior.

Calhoun's first argument seems contradictory, as the 1850 Fugitive Slave Act showed the South had no qualms about using the federal government to force free states to aid slave states in recovering their "property," and his second argument seems peculiarly tone-deaf, reminiscent of when a White Southerner encountered a Black in Canada who had escaped from slavery, and asked, "Did your master beat you, or mistreat you or your loved ones?"

The Black said, "No, Sir."

"Well, did he give you good food, clothes, and housing?"

"Yes, Sir."

"Well then, don't you regret running off to Canada where the winters are cold and the work is uncertain, away from the warm and sunny South where your master treated you like family, and fed you and clothed you and housed you for free?"

"Sir," said the Black, "I understand the position is available, if you are interested."

But the South had become addicted to slavery, which offered "Massa" a particularly heady drug of financial abundance. Slave-owning was an immensely profitable and growing business. At the time the South seceded in 1860-61, a slave was worth perhaps $150,000 or more in 2016 dollars, twice the average of 14 years earlier. A holder of 5-10 slaves enjoyed relative earnings of $4 million and economic power of $37 million; a holder of 500 or more slaves enjoyed relative earnings of $386 million and economic power of $3.93 billion in 2016 dollars. Every slaveholder probably ranked in the top 1% of the population. Anyone firmly for slavery was *sound on the goose*: staunchly protecting the goose that laid the golden eggs. In addition, a slaveholder wielded a physical, sexual, emotional, and mental dominion—even an almost godlike, absolute power of life and death—over other human beings.[11]

---

[11] "To be sound on the goose. *Verb. Phr.* (American). – Before the civil war, to be sound on the pro-slavery question: now, to be generally staunch on party matters; to be politically orthodox." John S. Farmer and W. E. Henley, *Slang and Its Analogues Past and Present: A Dictionary, Historical and Comparative, of the Heterodox Speech of all Classes of Society for More than Three Hundred Years*, Vol. 3, [London: n.p.], 1893, p. 182. For an extensive article on the value of slavery, see Samuel H. Williamson and Louis P. Cain, "Measuring Slavery in 2016 dollars," https://www.measuringworth.com/slavery.php.

Slavery was, after all, an American tradition. From 1619, when English colonists first brought African slaves to Virginia, Americans had depended heavily on slave labor. Even New Englanders had owned slaves. Massachusetts finally abolished slavery in 1783, but racism still thrived, and some New Englanders got rich buying slaves in Africa and shipping them enchained and horrifically close-packed to Caribbean sugar plantations. Often one in four slaves died en route. From there the merchants brought molasses back to New England to make rum, and then traded or sold the rum to buy more slaves in Africa. The last of the northern states to abolish slavery was New Jersey in 1804, and even then with gradual emancipation. New Jerseyans still held as many as 400 slaves at the close of the Civil War.[12]

Unlike in the West Indies, Southern Americans did not generally need a great many slaves until the White, Massachusetts-born Eli Whitney patented the cotton gin in 1794. The gin easily removed cotton seeds from bolls and made large cotton plantations lucrative, and suddenly-wealthy Southerners now needed lots of slaves to work their plantations. "King Cotton" also enriched the Northerners who built and ran the factories to turn slave-grown cotton into cloth. Few questioned the system. To put it bluntly, America was built on slavery; slaves made money, and money was a sign of divine favor. The misery of the enslaved was a necessary evil apparently endorsed by both the Bible and the U. S. Constitution.[13]

---

[12] The New England "triangle trade" in slaves, molasses, and rum was not as monolithic or extensive as has been claimed; after years of researching New England shipping records, historian Clifford K. Shipton couldn't recall "a single example of a ship engaged in such a triangular trade," which would have taken a year to complete. And although a ship might complete one or two legs of the triangle, "the value of the rum-for-slaves trade was minimal. It didn't come close to providing an economic engine for early New England." Wayne Curtis, *And a Bottle of Rum: A History of the New World in Ten Cocktails*, New York: Three Rivers Press, 2007, pp. 126-127.
Figures commonly cited for New Jersey slaves are 18 at the start of the Civil War and 16 at its close; Professor James Gigantino raises that last number to as many as 400: James J. Gigantino II, *The Ragged Road to Abolition: Slavery and Freedom in New Jersey, 1775-1865*, Philadelphia: University of Pennsylvania Press, 2015.

[13] Slave-owners often quoted Ephesians 6:5, Colossians 3:22, Titus 2:9-10, and 1 Peter 2:18, all urging servants to obey their masters. The U. S. Constitution, Article 1, Section 2, apportioned a state's number of representatives by its population, excluding all untaxed Indians and counting all those not "free" or "bound to service for a term of years"—i.e., slaves—as three-fifths of a person. This was modified in 1868 by the 14th Amendment, Section 2.

Slave Rebellion and Suppression

When John Brown planned and trained his men in Iowa for his Harper's Ferry slave rebellion in 1859, he would be confronting Southern slaveholders with their greatest nightmare. Slaves bloodily and successfully overthrew their French "masters" in St. Dominque in 1791-93, eventually creating the new nation of Haiti in 1804. Slaves also rebelled frequently in the States, including the Stono (or Cato's) Rebellion in South Carolina in 1739, the New York City conspiracy in 1741, Gabriel's conspiracy in Virginia in 1800, the German Coast uprising in Louisiana in 1811, Denmark Vesey's revolt in South Carolina in 1822, and Nat Turner's rebellion in Virginia in 1831. And on Christmas Day of 1856, the anti-slavery *Fairfield Ledger* carried this news, datelined Washington, Dec. 15:

> The people of Alexandria, Va., have been greatly alarmed for several days past in consequence of a threatened insurrection among the negroes. The military were called out last night, and 22 slaves were arrested at a ball where they had assembled without permission, against the laws of the State.
>
> Governor Wise had supplied arms and ammunition, and the people throughout the county of Alexandria and Fairfield [*sic*] are arming themselves in case of a general outbreak.
>
> No evidence has been found against any of the slaves arrested.[14]

Another *Ledger* column, datelined Louisville, December 17, showed slave-rebellion was threatening not just Virginia, but also Kentucky:

> The correspondent of the Journal writing from Campbellsville, Ky., on the 10th says that a negro boy had disclosed a plot of the negroes in that neighborhood to rise about Christmas, several arrests had been made, and an examination held on the 9th, before Justice Cloyde but nothing elicited save the statement of the boy, that he overheard the negroes say that they intended to make war upon the whites about Christmas, and that if he would join them they would make him rich. The negroes are reported to possess guns, pistols, &c. The correspondent adds that considerable dissatisfaction exists generally

---

[14] *Fairfield Ledger*, Dec. 25, 1856, p. 2, col. 3. "Fairfield" was probably an error for "Fairfax," the county adjoining Alexandria. Gov. Wise would order the hanging of John Brown for his own unsuccessful slave uprising at Harper's Ferry, (West) Virginia in 1859.

among the negroes which if not promptly suppressed may lead to serious trouble.[15]

Suppression was the watchword. The ever-present and terrifying threat of slave rebellion required the master to keep his "property" in powerless ignorance, like livestock. Slaves could not legally marry, own property, learn to read or write, stay out after curfew, assemble in groups without a White, or testify in court against a White. The master always carried a heavy cane and Bowie knife or pistol. "Uppity" Blacks were beaten or whipped, and "incorrigible" Blacks were shot, hanged, or sold downriver into hard labor in the cotton fields of Alabama or Mississippi, where life was brutal and short, and escape nearly impossible. Anyone who encouraged slaves to rebel or escape, or who even questioned the system, was a Black-loving abolitionist: an anarchist, terrorist and thief, liable himself to be beaten, hanged, or imprisoned at hard labor.[16]

With a breathtakingly circular logic, the master could now argue that the Blacks he had so disempowered and dispirited were inherently docile, servile, content, and loyal, incapable of taking care of themselves. He hid his feudal tyranny and violence behind the appealing roles of the beneficent *paterfamilias* shouldering the "White man's burden" of English imperialism, and the chivalrous medieval knight with his loyal serfs from the romantic novels of Sir Walter Scott, so popular in the South. Even the poorest Whites, those with no hope of ever owning slaves themselves, now had an underclass they could comfortably look down upon and use or abuse.

Thus enthralled, many Southern Whites found Calhoun's arguments convincing, so that by the 1840s and '50s most supported "states' rights"—meaning "Southern states' rights to enforce slavery even in the North, and to secede if those 'rights' weren't met"—and echoed Calhoun's claim that slavery was a "positive good" to be perpetuated always and spread everywhere. Most Northerners wouldn't go that far, but they originally agreed, at least reluctantly, that abolitionists were traitors fomenting disunion, and perhaps even that Underground Rail-

---

[15] Ibid., p. 2, col. 6.
[16] For American slave rebellions, see Henry Louis Gates, Jr., "Did African-American Slaves Rebel?" online at PBS, *The African Americans: Many Rivers to Cross*, http://www.pbs.org/wnet/african-americans-many-rivers-to-cross/history/did-african-american-slaves-rebel/

road operators were both traitors and thieves, disrupting the delicate balance between North and South and stealing valuable property.[17]

The Roots of Discord between North and South

The antebellum conflict between the North and South began about 200 years earlier, almost from when England first colonized America. A great many Puritans from East Anglia emigrated for Massachusetts between 1629 and 1641, to escape Anglican and royal displeasure. When King Charles I attempted to arrest leading members of Parliament in 1642, the Puritan and Parliamentarian "Roundheads" began the Civil War with Charles's Royalist "Cavaliers," and displaced many elite Cavaliers from the south of England, who emigrated for Virginia with their indentured servants.[18]

The Roundheads favored a more-or-less democratic government, and were mostly Presbyterian or Puritan, sober, middle-class, mercantile townsfolk who considered labor to be godly. The Cavaliers favored an autocratic royal and feudal government, and were mostly Catholic or Anglican, sensual, agrarian plantation-folk who disdained labor and delegated it to their servants and serfs.

In later centuries, the Roundheads became Whigs; the Cavaliers became Tories and later Democrats, but their earliest attitudes persisted in America in religion, politics, character, and even personal appearance. New England gentlemen were still mostly Presbyterian or Congregationalist, and still favored the Roundheads' urban and mercantile middle-class democracy. They maintained a sober lifestyle and garb with short hair, avoiding the appearance of luxury, and used the courts to

---

[17] For the psychology of slavery, see William W. Freehling, *The Road to Disunion*, vol. I: *Secessionists at Bay, 1776-1854*, New York: Oxford University Press, 1990, and *The Road to Disunion*, vol. II: *Secessionists Triumphant, 1854-1861*, New York: Oxford University Press, 2007, and Scott Horton, "How Walter Scott Started the American Civil War," *Harper's Magazine*, July, 2007, online at https://harpers.org/blog/2007/07/how-walter-scott-started-the-american-civil-war/. For a clear exposition of the change in Iowa's antislavery attitude from escapism to engagement, see James Connor, "The Antislavery Movement in Iowa," *Annals of Iowa*, vol. 40, no. 5 (Summer 1970), pp. 343-376, and vol. 40, no. 6 (Fall 1970), pp. 450-479, http://ir.uiowa.edu/annals-of-iowa/vol40/iss5/5/; http://ir.uiowa.edu/annals-of-iowa/vol40/iss6/6/.

[18] See David Hackett Fisher, *Albion's Seed: Four British Folkways in America*, New York: Oxford University Press, 1989.

correct injustice. On the other hand, Southern gentlemen were still mostly Catholic or Episcopalian, and still favored the Cavaliers' agrarian autocratic elitism, exuberant lifestyle and garb with long hair, often with a Van Dyke beard like King Charles I. They traveled always with horse and sword, and to correct injustice they used personal force through a "code of honor" requiring duels between equals, and flogging or caning of inferiors. Sir Walter Scott's romantic novels of medieval chivalry did not introduce new ideals to the South, but confirmed the age-old ones it already held.

During and after the American Revolution, the Articles of Confederation and the Constitution patched over the radical differences between the northern and southern states with compromises, but perhaps Lord North was correct in calling the American Union "a rope of sand." Certainly the seeds of disunion sown in the English Civil War bore bitter fruit two centuries later in the American Civil War.

The nationwide embrace of Victorianism about 1840 muted some of the differences between North and South in appearance, but not in core values. North and South remained profoundly at odds in their attitudes towards serfs or slaves.

Early Attitudes in Iowa

Originally, the North's hands-off but racist attitude was shared by most Iowans, including Jefferson County settlers. Iowa was established in 1838 as a free territory and brought into the Union in 1846 as a free state, but the first legislators of Iowa Territory wanted to discourage free Blacks from immigrating. They had quickly barred Blacks from attending free public schools, voting, performing military service, intermarrying with Whites, or bearing witness against a White in court. Anti-slavery constituents in Denmark and Salem quickly protested the Black laws, but to no avail. Worse still, Iowa's 1840 census listed 16 slaves in Dubuque County. General Joseph M. Street of Virginia, Wapello County's Indian agent, openly kept slaves, whom he called "indentured servants." Although Jefferson County had some antislavery settlers, it also had many pro-slavery or at least pro-Southern settlers, and a few also kept several Blacks in slavery.[19]

---

[19] Robert R. Dykstra, "White Men, Black Laws," *Annals of Iowa*, vol. 40, no. 6 (Fall 1982), pp. 403-440, http://ir.uiowa.edu/annals-of-iowa/vol46/iss6/2/. For the 16

Pro-slavery Currents in Jefferson County

On the pro-slavery side, Fairfield was only 30 miles north of the slave state of Missouri, and had strong business and familial ties there. The Burlington & Missouri River Railroad did not reach west from Chicago to Fairfield until September 1858. The Midwest's major antebellum trade route was the Mississippi River, and Fairfield's nearest metropolis was St. Louis. It boasted a population of nearly 78,000 in 1850, when Chicago still had less than 30,000. St. Louis businesses regularly advertised in Fairfield newspapers; Chicago businesses did not.

In 1838, Henry County's Democratic Representative William G. Coop of Walnut Creek pushed for a separate county—Jefferson—with his own town of Lockridge as its county seat. Henry County's Democratic Senators Lawson B. Hughes and Dr. Jesse D. Payne opposed the split, as it would make Jefferson County mostly Democratic but leave Henry County mostly Whig. Nonetheless, on January 21, 1839, Col. Coop got his county, though in March, to his disappointment, the commissioners located its seat west of Lockridge nearer the center of the county: in "Jefferson Court-house," renamed "Fairfield" that May.[20]

Although from its beginning Fairfield enjoyed an unusually diverse blend of manufacturers and professionals, Jefferson County was still predominantly a Democratic farming community, in many ways more in tune with the agricultural South and with most of the United States, than with the federalizing Whigs of Henry County and the industrial North. The county was named for Thomas Jefferson, the ultimate Virginia Democrat, who while abhorring slavery and striving to prohibit it from the Northwest Territory, kept hundreds of slaves himself and

---

slaves in Dubuque, see "Slaves in Iowa," *Annals of Iowa*, vol. 6, no. 1 (1903), p. 66, http://ir.uiowa.edu/annals-of-iowa/vol6/iss1/13/ Iowa's territorial law banning interracial marriage was eliminated relatively early, in 1851. Iowa first admitted Blacks into the military during the Civil War, after the Emancipation Proclamation in 1863: David Brodnax, Sr., "Will They Fight? Ask the Enemy: Iowa's African American Regiment in the Civil War," *Annals of Iowa*, vol. 66, no. 3 (Summer 2007), pp. 266-292, http://ir.uiowa.edu/annals-of-iowa/vol66/iss3/3/. Iowa's public schools were integrated in 1868, with the Supreme Court decision in *Clark v. Board of Directors* allowing African-American Susan Clark to attend public school in Muscatine. Iowa was also the only Northern state whose voters extended suffrage to Blacks, giving Blacks the right to vote in 1868: David Brodnax Sr., op. cit., p. 290, n. 72.

[20] *The History of Jefferson County, Iowa, Containing a History of the County, its Cities, Towns, &c....*, Chicago: Western Historical Co., 1879, p. 377.

thought emancipation must be democratically chosen by the slave-owners. The only slave family he himself ever freed were the children he as a widower reportedly sired with his wife's half-sister, Sally Hemings.

Slaves in Fairfield

Antebellum Fairfield was actually home to several slaves, though technically illegal in the free Territory and State of Iowa. Fairfield's first slave was a 12-year-old Black boy, bought in Kentucky by John Ratliff, who sold groceries and liquor at 61 East Broadway. The boy wished to become a jockey, and he was soon sold for $200 to a Mr. Dill, who owned some fast horses, presumably in the South. Ratliff would have owned and sold "his" slave probably before he sold his store in 1842, and of course before he died in 1846.[21]

Fairfield's second slave was young William Henry Harrison Triplett, kept in slavery in Fairfield from about 1846 until as late as 1854. He was born in 1833 in Ft. Crawford, Wisconsin to "Aunt Patsy," a nearly-White Quadroon, and a Virginia slave who was part Indian. They were slaves who had been "'indentured'… far in excess of Territorial law's allowance of 19 years" to the slaveholding Virginian General Joseph M. Street. The general gave the two-year-old Henry to his daughter Mary on her marriage to Captain George Wilson, a slaveholder from Ohio who sometime in the late 1830s sold an "unruly" slave for a pair of mules. Leaving young Henry with the Wilsons in Wisconsin, General Street took the rest of Henry's family in 1839 to Iowa, where he was the Indian agent of Agency. Here he sold several slave women purchased in Missouri, and he freed Charles Forrester and the other

---

[21] For John Ratliff's slave, see *The Fairfield Ledger*, Oct. 2, 1939, p. C-5, col. 3. Perhaps Mr. Dill was related to Stephen Dill of Tennessee and Fairfield, whose daughter Elizabeth married John W. DuBois. In 1866, 20 years after John Ratliff died, his heirs asserted that in 1842 John had feared his alcoholism was leading him to ruin and so had signed his property at 53-61 East Broadway over to Parish Ellis in trust for Ratliff and his family. As Parish or his widow Mehitable had now sold all the old Ratliff land but the corner at 61 E. Broadway and 100-110 N. Court, the heirs petitioned to reclaim it but could not prove that Ratliff's deed to Parish Ellis was not absolute. That summer Mehitable sold the corner property to George W. Vance for $3000: WD dated July 26 and filed Aug. 4, 1866 in Jefferson Co. Deed Records Book 4, p. 231.

Tripletts by will at his death in 1840, but Henry was to obtain his freedom only at age 21 in 1854, after being taught a trade.[22]

By 1844 George Wilson was surveying land just west of Fairfield, Iowa for townships. He wrote from "Wapello County, Iowa, near Fairfield" on January 23 to Surveyor General James Wilson in Dubuque, concluding, "Address me at Fairfield, Jefferson County and please consider me an applicant for further employment," but he does not appear to have moved to Fairfield permanently at the time.[23]

Henry Triplett finally came to Iowa with the Wilsons in 1846; Captain Wilson's son later recollected:

> When we lived on the farm near Agency the boy Henry saved my mother many a step and was a great help, as he was bright, active and intelligent. Gen. Street's sons took up a claim for Patsy and had to lie out with their rifles more than one night to defend it from "jumpers." Gen. Street set all his slaves free by his will. Henry was to be kept by my father until of age, and before his majority was to be taught a trade. When he was a little boy, he says, he had a great admiration for his master, Wilson, and tried especially to hold himself straight and walk like him.... On a visit ... he said, "I used to think your father was too hard on me and you boys when we were little, but I have many a time thanked him in my mind for his rigid discipline and training." On one of his visits... he said, "Get out your father's old leather-covered family Bible that he always used in family prayers when we were boys. I would like to have prayers with you all before I go, as I have so many times listened to my faithful friend and master

---

[22] For Henry H. Triplett, see Leslie A. Schwalm, *Emancipation's Diaspora*, Chapel Hill: University of North Carolina Press, 2009, pp. 18-19.

[23] Letter offered for sale by "99counties," Civil War Americana Antiques, in April 2018 on eBay. Addressed to "Genl. James Wilson, Surveyor Genl., Dubuque," he letter reads: "Sir, I have the honor to transmit herewith the Field Notes of the Townships of my contract.

"I have met with many hindrances in the execution of my work, chiefly from the exceeding roughness of the county; but also from sickness among my party, caused by using the very bad water found in these Townships.

"I have bestowed the utmost pains upon the survey, and hope it will prove acceptable.

"Address me at Fairfield, Jefferson County and please consider me an applicant for further employment. Very respectfully Your obt. Servt. Geo. Wilson"

reading from it." He read and prayed fervently and it seemed to do him much good, and was a gratifying circumstance to all present.[24]

The 1850 U. S. census shows Wm. H. Triplett in Fairfield still living with land-office register George Wilson, wife Mary, and their six children. Henry's race and occupation were left blank. Captain Wilson's dislike of slavery reportedly grew, though not because of its cruelty; he saw it as humane, but because it "made the slave-holders an overbearing, tyrannical class." He evidently did free Henry in 1854. Two years later the state census shows that Henry had reunited with his family in Agency. He was a blacksmith in Keokuk in 1860, living with wife "Candes" and stepchildren, all of Tennessee, as well as his mother.

During the Civil War, William H. Triplett enlisted on September 10, 1861 in the 2nd Iowa Regimental band as 3rd class musician, mustering out May 1, 1862 at Pittsburg Landing. Henry was a blacksmith in Missouri in the 1870s and '80s, living with his wife Kate of Tennessee and their children, all born in Missouri. His mother still lived with them. He later became a minister of the African Methodist Episcopal Church, and was "a man of high character and of self-respect, spoken of in the most approving terms by those who know him …, and a faithful and intelligent worker for the moral elevation of his people."[25]

---

[24] George Wilson (Jr.), "George Wilson: First Territorial Adjutant of the Militia of Iowa," *Annals of Iowa,* vol. 4, no. 8 (Jan. 1901), pp. 571-574, http://ir.uiowa.edu/annals-of-iowa/vol4/iss8/2/.

[25] The 1850 federal census lists Wm. H. Triplet, 18, b. Va., last in the household (99-101) of land-office register George Wilson. His occupation was blank; next below was the household (100-102) of blacksmith Ruben Fenstermaker (who was town marshal of Farmington when shot and killed in 1867 by a suspected horse-thief). The 1856 Iowa census for Agency, Wapello Co., shows Henry, 23, in the household (76-76) of his widowed mother Martha, 56, b. in Ky., with brothers London, 26, Lewis, 21, and J. N., 19. All were 17-year residents of Iowa except Henry, resident for 10 years. None had an occupation except "Lewis," a blacksmith: probably actually Henry. The 1860 U. S. census for Keokuk, Lee Co., Iowa shows (813-763) Henry Triplett, 27, as a Wisconsin-born blacksmith in Keokuk with $200 in personal estate. Living with him were his wife Candes, 28, his mother Martha, 60, and William T., 8, and Alonzo Sweedes Triplett, 11. Like Candes, the children were b. in Tenn. and probably his stepchildren. In 1880 he was 44, a blacksmith on Lilly Street in Chillicothe, Livingston Co., Missouri (43-101); with him were wife Kate, 36, b. in Tenn., and children Henry, 12, Edgar, 10, May, 7, Isiah, 5, James, 2, all b. in Missouri, and mother Martha, 80, b. in Va. For Wm. H. Triplett as regimental musician in Civil War: *Roster and Records of Iowa Soldiers in the War of the Rebellion, Together with*

Fairfield's third slave remained a captive here only briefly. In the late spring of 1862—about a year into the Civil War—David F. Phillips, claiming to be an active-duty surgeon of the 43rd Ohio regiment, was staying with Fairfield photographer Moses C. Shamp and had with him a Black "servant" boy named James Robinson. Conflicting stories sparked the suspicion that James was a slave, and Fairfield's abolitionist blacksmith Melchi Scott and his Republican lawyer A. M. Scott obtained a writ of *habeas corpus* on June 10 from County Judge A. R. Fulton. Democratic attorneys Negus & Culbertson defended M. C. Shamp and Dr. Phillips.

The trial on June 11 revealed that although Phillips had told James to say he was born and raised in Gallion, Crawford County, Ohio "to quiet the apprehensions of the 'Abolitionists,'" James was actually a slave in Missouri given to Phillips, who was keeping him in custody and had brought him voluntarily into Iowa, where slavery was forbidden. Phillips claimed he was keeping James as his servant only so long as he was a surgeon in the army and had need of him in that capacity; afterwards he would release him and find him a suitable position. Judge Fulton declared James to be free; Shamp and Phillips appealed the decision, but in vain. A witness added that the "braggart, bullying and ungentlemanly conduct of Phillips and his attorney was in perfect keeping with the whole case."[26]

The regimental roster shows that 2nd Lieutenant—not Surgeon—David F. Phillips had actually resigned from the 43rd Ohio Infantry on May 17, 1862. Dr. Phillips was Moses C. Shamp's son-in-law, having married his daughter "Artamesia Desdemonia" in 1844.[27]

Historical Sketches of Volunteer Organizations, 1861-1866, Des Moines: E. H. English, 1908-1911. vol. 1, p. 100, http://iagenweb.org/civilwar/books/logan/mil302.htm. Quotes by George Wilson (Jr), op. cit., pp. 573-574, http://ir.uiowa.edu/annals-of-iowa/vol4/iss8/2/.

[26] *Fairfield Ledger*, June 26, 1862, p. 2, col. 4; Charles J. Fulton, *History of Jefferson County, Iowa: A Record of Settlement, Organization, Progress and Achievement*, Chicago: S. J. Clarke Publishing Co., 1912-1914., vol. 1, p. 354. Fulton names the "servant" Ralph Robinson.

[27] Charles A. Poland, *Army Register of Ohio Volunteers in the Service of the United States ... for July, 1862*, Columbus, Ohio: Ohio State Journal Pub. Co., 1862, p. 80, https://archive.org/stream/armyregisterofoh00pola#page/80. For Phillips as Moses C. Shamp's son-in-law, see Find A Grave: Artamesia Desdemonia *Shamp* Phillips, David F. Phillips: https://www.findagrave.com/memorial/76463844/artamesia-desdemonia-phillips; https://www.findagrave.com/memorial/76463920/david-f-phillips

Prominent Pro-slavery Fairfielders

As might be expected, many of Fairfield's pro-slavery settlers came from Tennessee, Kentucky, and Virginia, and the lawyer and land-agent William L. Hamilton was from Georgia, but among the town's most prominent pro-Southern Democrats, the so-called "land-office clique," both Hon. Bernhardt Henn (Fairfield's land-office register in 1846-49) and Col. James Thompson (land-office register in 1853-55) were born in New York. Hon. Charles Negus was from Massachusetts, though marrying into a slaveholding family from Virginia. Both Hon. Samuel H. Bradley and Dr. Jesse C. Ware were from Pennsylvania, as was the *Jeffersonian* editor Samuel Jacobs, and the *Sentinel* editor David Sheward came from Ohio. As mentioned, slaveholder Capt. George Wilson (land-office register in 1849-51) was also from Ohio, though also marrying into a Virginian slave-holding family. The strongly pro-slavery merchant Michael M. Bleakmore, later editor and publisher of *The Iowa Democrat*, was from (West) Virginia.

Anti-slavery currents in Jefferson County

On the anti-slavery side, the Congregationalists of Denmark created Iowa's first Anti-Slavery Society on Jan. 1, 1840, just after they had pushed west to establish a church in Fairfield on December 21, 1839. The Quakers of Salem created the Salem Anti-Slavery Society in February 1841, and they moved west and north to establish meetings at Pleasant Plain in 1841, and at Richland in 1851. Anti-slavery Presbyterians established settlements just south of Jefferson County in Van Buren County's Kilbourn in 1841 and Birmingham in 1844, and just north of Jefferson County in Washington County's Brighton and Washington about the same time.

There appears to have been no anti-slavery society in Jefferson County. It was predominantly Democratic, and most of its population was "Western" and viewed "Yankees" with suspicion. Early UGRR operators in Jefferson County kept their activities private and secret.

Jefferson County's anti-slavery activists and UGRR operators eventually included Congregationalists, Quakers, Methodists, ex-Quaker Methodists, Presbyterians, ex-Presbyterian Congregationalists, and the occasional free-thinking Unitarian or Transcendentalist. While most of these came from New England, the Midwest (often southwestern Ohio), and the Mid-Atlantic States (often Washington County, Pennsylvania), many of the Quakers had fled the slave-culture of North Carolina, and William Wallace Junkin, the strongly anti-slavery *Fairfield Ledger* editor and publisher, was from (West) Virginia.

1839-1846: Anti-slavery and a probable UGRR in Fairfield

Before the Civil War, pro-slavery Fairfielders appeared always willing to uphold their interpretation of the Bible and the Constitution, obey the federal law, and help the posses of armed and angry Missouri slave-catchers spy on suspected UGRR households and recover "their" escapees. The strong pro-slavery presence probably necessitated more UGRR secrecy in Fairfield than in the more homogenous Denmark or Salem. Even in the relatively Whiggish Mount Pleasant nearby, the heroic UGRR station master Samuel L. Howe was "bitterly hated" as a Free-Soiler and vocal abolitionist, and "had to endure the wickedest persecution:"

> His property was destroyed, his stock stolen, emissaries were sent to take his life, and finally he was brutally mobbed by the pro-slavery ruffians in the streets of Mt. Pleasant.... He defied persecution, hatred, loss of property, social ostracism and even dared death itself in defense of those immortal principles that afterwards became the chief corner-stone of the great National Republican Party.[28]

Fairfielders may have established an Underground Railroad network as early as 1839, the year the town was founded. That December 21-22, Rev. Reuben Gaylord of Connecticut organized the Congregational Church in Fairfield. The minister and his congregation all appear to have been strongly anti-slavery. Before graduating from the Yale Theological Seminary, Rev. Gaylord had taught and studied theology at

---

[28] W. P. Howe, "Fifty-Two Years in Iowa," *The Annals of Iowa*, vol. 1 (1894), p. 570. Samuel L. Howe was a UGRR station master in Mt. Pleasant: "After the destruction of the Whig party my father became a free soiler.... At a very early period his house was a station on the 'underground railroad' (the first one north of Salem), and many a poor colored man, woman and child did he help on their way to freedom, bidding them God speed…with transportation, money, food and clothing." *Loc. cit.* http://ir.uiowa.edu/annals-of-iowa/vol1/iss7/7

Illinois College in Jacksonville in 1834-37 under its abolitionist president, Rev. Edward Beecher, brother of Henry Ward Beecher and Harriet Beecher Stowe. President Beecher befriended the abolitionist printer Elijah P. Lovejoy, and wrote a moving account of his murder by a pro-slavery mob in 1837, and managed the Illinois Anti-Slavery Society the following year. Founded in 1829, Beecher's college "had close ties with the abolitionist movement and the Underground Railroad from its inception."[29]

All of the Fairfield Congregational Church's twelve original members had ties to anti-slavery, abolitionism or the UGRR. Deborah (Murray) and Dr. Jeremiah Waugh of Ohio founded a temperance hotel called the Restoration House on the west side of the square in 1840, soon after Jeremiah's anti-slavery letter appeared in the abolitionist Theodore Weld's 1839 *American Slavery As It Is: Testimony of a Thousand Witnesses*.[30]

To teach their children that first winter of 1840-41, Jeremiah and Deborah hired Clarissa Sawyer of Denmark, Iowa, where her parents belonged to the Congregational Church of UGRR manager Rev. Asa Turner; she herself would join Rev. Turner's church in 1842. She taught in an upper room of the Restoration House. Also attending Clarissa's school was the son of a third founding member, storekeeper Ebenezer Sumner Gage of Maine, who was elected the church's first clerk and treasurer. Ebenezer was an early member of the anti-slavery Republican Party.[31]

The church's other nine charter members were the parents, siblings, and in-laws of George B. Hitchcock of Massachusetts, later famed for his UGRR house near Lewis, Iowa. George himself would soon come from Jacksonville, Illinois to Fairfield, and study theology under the abolitionist Julius A. Reed of Connecticut, the Fairfield Congregational Church's first minister. Rev. Reed had come to Fairfield in 1840 at the

---

[29] Owen W. Muelder, *The Underground Railroad in Western Illinois*, Jefferson, North Carolina: McFarland & Co., Inc., 2008, p. 165n58; National Park Service, U. S. Dept. of the Interior, "Aboard the Underground Railroad: Beecher Hall," https://www.nps.gov/nr/travel/underground/Beecher_hall.html.

[30] Theodore Dwight Weld *American Slavery As It Is: Testimony of a Thousand Witnesses*, New York: American Anti-Slavery Society, 1839, p. 177.

[31] Clarissa Sawyer joined Asa Turner's church on Sept. 4, 1842: Barbara Scott, "Denmark Congregational Church Records, Denmark Township., Lee County, Iowa," Lee County Iowa GenWeb Archives, online at http://files.usgwarchives.net/ia/lee/churches/dmkchr.txt

request of his old friend Rev. Asa Turner of Massachusetts, the founder of the UGRR network in Denmark, Iowa. In his letter inviting Rev. Reed to Fairfield, Rev. Turner mentioned the "Yankee settlement" of Brighton about twelve miles north and told him to "enlist" Allen B. Hitchcock, George's brother, "for Iowa."

Before reinforcements arrived, Rev. Julius Reed divided his time between Fairfield, Brighton, and Keosauqua. A logical UGRR route from Missouri would lead ten miles north to cross the Des Moines River at Keosauqua, then 20 miles north to Fairfield, then about 16 miles north-northeast to Brighton.[32]

George B. Hitchcock's brother Allen settled in Davenport, where he built the first Congregational church and served as its minister in 1841-43. George and Allen's brother, Jared Beecher Hitchcock, was the Davenport church's first deacon. Davenport was a natural eastern terminus for the Iowa UGRR, as fugitives could cross the Mississippi here to Rock Island or to Moline, Illinois, where Allen moved next, leaving his brother Jared in Davenport, and its church to Rev. Ephraim Adams, one of Asa Turner's eleven "Iowa Band" abolitionist missionaries from Massachusetts. Once in northern Illinois, a freedom-seeker could easily move on to relative safety in Chicago, and ultimate safety in Canada.[33]

---

[32] On Feb. 25, 1841, Rev. Reed reported to Rev. Milton Badger that he intended to preach at Fairfield two Sabbaths out of every three, every sixth Sabbath in Washington County to the north, and the other Sabbath at Keosauqua. On May 28, 1841, Reed reported that over the last quarter he had labored at Fairfield two-thirds of the time and temporarily at Keosauqua one-third of the time. On Dec. 2, 1841 he reported that he had helped form churches at Brighton and Keosauqua, and had assisted Bro. Gaylord. State Historical Society of Iowa, MSS: .R251 Rev. Julius A. Reed, Correspondence. UGRR stations soon arose midway between Keosauqua and Fairfield at Birmingham, where UGRR operator and Presbyterian Rev. David Lindsay settled by 1844, and 12 miles NNE of Fairfield at Pleasant Plain.

[33] The "Iowa Band" was a group of 11 pioneer ministers from Andover Theological Seminary in Newton, Massachusetts. (A 12th, William B. Hammond, was detained by illness and, fearing the Western climate, did not come to Iowa.) In 1842, 5 of the 11 had heard Lyman Beecher prophecy in Oberlin that if America was to lead the world in "moral and political emancipation" it was now time "she understood her high calling, and was harnessed for the work:" Her "religious and political destiny" would be decided in the West, and hinged upon Iowa: There was the territory, and there would soon be the "population, the wealth, and the political power." And it must be done quickly. "Mighty influences are bearing on us and a slight effort now may secure what ages of repentance cannot remove away. We must reveal the Kingdom to the whole nation while we may. All who are alive must be enlightened, and reached by

Three months after Asa Turner's Denmark Congregationalists founded Iowa's first Anti-Slavery Society on Jan. 1, 1840, the anti-slavery Liberty Party held its first National Convention at Albany, New York on April, 1, 1840. It had split from the American Anti-Slavery Society in 1839, six years after the society's founding, to pursue its goals through politics. Asa must have welcomed the anti-slavery movement's political debut; he worked tirelessly over the coming years to form anti-slavery political coalitions which from very modest beginnings gradually acquired an inexorable momentum. The Liberty Party folded in 1848 into the Free-Soil Party, which in 1854-56 formed the core of the Republican Party, whose early success in Iowa owed much to Asa Turner's efforts.

After Asa Turner, Reuben Gaylord, and Julius Reed founded the Congregational Association in Denmark in November 1840, Asa Turner moderated the Association's second meeting in April 1841 in Fairfield, where "they adopted resolutions against slavery as thorough as any that were adopted in later years." That year Julius Reed gave the principal speech in Fairfield at the founding of the County Temperance Society, with George B. Hitchcock serving as his vice president. According to Rev. Reed, all of the Congregationalist ministers in Iowa were abolitionists, preaching constantly for thirty years on the need for both temperance and freedom.[34]

---

the restraining and preserving energies of Heaven." The band came in 1843 to Iowa, where they established Iowa College and Grinnell College, along with more than 200 schools and Bible classes, and over 600 churches. See Greg Crawford, *Iowa's Prophetic Journey: Four Anointings*, CreateSpace, 2012, p. 11; Ephraim Adams, *The Iowa Band*, Chicago: The Pilgrim Press, n.d., ca. 1902.

[34] "There may have been differences of opinion on other subjects… but these… could wait and at all events must not stand in the way of the establishment of temperance and freedom.

"The ministers and delegates who organized the Association of Iowa … were all of them avowed abolitionists, and the zeal of some of them had been kindled anew by the … murder of Lovejoy. At their second meeting, held in April 1841 at Fairfield, they adopted resolutions against slavery as thorough as any that were adopted in later years. For 30 years there was no Congregationalist minister in Iowa who did not practice and advocate total abstinence from all that can intoxicate, and did not also advocate the abolition of slavery.

"Frequently temperance and slavery were the subjects of sermons, and a religious service was rarely held in which one or both of these subjects were not mentioned." Julius A. Reed, *Copies of Historical Papers and Addresses*, p. 216, cited in Leah Rogers, National Historic Landmark Nomination of the Reverend George B. Hitch-

George B. Hitchcock was the deacon of Rev. Reed's Fairfield church in 1844, just before embarking on his own ministries in Oskaloosa, and in Eddyville, and—after exploring the area in 1850 with Rev. Julius A. Reed—near Lewis, Iowa: all towns that afterwards became known UGRR stations, in his wake.

Several other residents of Fairfield in the early 1840s, including Thomas Mitchell and his friend Herbert M. Hoxie, also ran known UGRR stations deeper in Iowa later, and it seems likely they all may have begun quietly in Fairfield. Pioneer Dr. John Jackman Smith of Libertyville, by contrast, had freed "his" slaves down South before coming to Iowa, and he may have continued his service on the UGRR here. But secrecy here was crucial, as Fairfield was not a "Yankee settlement." As late as July 29, 1845, Rev. Julius A. Reed wrote:

> I think my influence is extending & my hold upon the people strengthening. This is a Western population & the prejudice against Yankees is strong & also against all anti-slavery feeling. Efforts have also been made I suppose to prejudice people against me. I have suffered much in my feelings. But these things are passing away, I have attended to my duties & have advanced against the current a little. But much hard work is yet to be done & much to be endured before much can be effected.[35]

By now, Iowans were considering statehood. Southern congressmen had finally agreed to allow Iowa as a free state if Florida was admitted as a slave state, which duly occurred on March 3, 1845. After much debate over state boundaries and a constitution, Iowa finally entered the Union as a free state on December 28, 1846. Iowa's new constitution prohibited public banks; the Panic of 1837 had spawned an ongoing recession, and many Southern and Western banks had irresponsibly issued paper money before failing, and ruining depositors.

The new state of Iowa had a Democratic governor, two Democratic U. S. congressmen, three Democratic justices of the state Supreme Court, and eventually, after much political wrangling, two Democratic

---

cock House, U.S. Dept. of the Interior, National Park Service, p. 24, http://focus.nps.gov/pdfhost/docs/NHLS/Text/77000500.pdf.

[35] George B. Hitchcock was deacon of Fairfield's Congregational Church in 1844: Organization of the Fairfield Congregational Church, dated Jan. 3, filed Jan. 16, 1844, Jefferson County Deed Book B, p. 350. Letter from Julius A. Reed to the American Home Missionary Society datelined Fairfield, Iowa, July 29, 1845, AHMS Archives, Amistad Research Center, Tulane University, New Orleans.

U. S. senators. But in the meantime, a day less than a year earlier, Texas had been admitted as another slave state, even though Mexico had not recognized its independence, precipitating the Mexican War of 1846-47.

1847: Anti-slavery and the UGRR in Fairfield

Probably partly because the pro-slavery aggression of the Mexican War distressed many Northerners, 1847 was a good year for anti-slavery in Fairfield. The Presbyterian saddler and harness-maker Joseph A. McKemey of Washington County, Pennsylvania reportedly met then with seven other Fairfielders and organized an Underground Railroad network here.[36]

Theirs was probably not Fairfield's first UGRR network, but it is the first for which we have an overt claim. Within that past year County Prosecutor George Acheson, a Democrat also from Washington County, Pennsylvania, had refused to arrest the freedman Charles Forrester, who had opened a barber shop in Fairfield without posting the $500 bond required by Iowa's racist Act to Regulate Blacks and Mulattoes. While Acheson's refusal was reasonable—Forrester had resided in Iowa longer than any of them, before the Act was created in 1839—still it shocked many Whites, and gave Blacks hope that they might have some friends in power here. With Fairfield now enjoying a population of about 650, it incorporated as a town or city in April 1847, and Fairfielders elected George Acheson their first city recorder, and Joseph A. McKemey a trustee. That September 15, seven Freemasons organized Fairfield's first lodge, and initiated first, George Acheson on the 16th, and third, Joseph A. McKemey on the 20th. Anti-slavery ties were augmented by familial ones that December 23 when George Acheson married Joseph A. McKemey's sister-in-law.

The Masons' lodge master was Dr. Jacob L. Myers, whose father-in-law had founded Fairfield's (Old School) Presbyterian Church, to which Joseph A. McKemey then belonged. While Dr. Myers and his in-laws were all from Fluvanna County, Virginia, he was among those

---

[36] *Portrait and Biographical Album of Jefferson and Van Buren Counties, Iowa...*, Chicago: Lake City Publishing Co., 1890, pp. 243-244; Charles J. Fulton, *History of Jefferson County, Iowa: A Record of Settlement, Organization, Progress and Achievement*, Chicago: S. J. Clarke Publishing Co., 1912-1914, Vol. II, p. 69.

calling for and participating in the People's Republican Convention in Fairfield on July 14, 1855. The lodge's senior warden was *Sentinel* editor Augustus R. Sparks, who on September 18, 1847 ran a front-page article favoring the Wilmot Proviso, a proposal to ban slavery from any territory acquired in the Mexican War. Junior Warden James Jeffers and Senior Deacon William P. Winner both also signed the call for a People's Republican Convention in 1855. Lodge Secretary Ebenezer Sumner Gage was a town trustee with Joseph A. McKemey in 1847, as well as a charter member of the anti-slavery Congregational Church; his name led the list in 1855 calling for a People's Republican Convention, followed immediately by Joseph A. McKemey's.[37]

Iowa Grand Master Ansel Humphreys of Connecticut and Muscatine officially installed J. L. Myers as master of Fairfield's Clinton Lodge, No. 15 on November 25, 1848 at Salem, Henry County, Iowa: about six months after citizens of that town famously defied Ruel Daggs' slave-catchers and helped "his" slaves escape from Missouri. That same November day, Grand Master Humphreys also organized the lodge at Salem under dispensation, and spent several days instructing, lecturing, and conferring degrees. Lawyer William M. Gordon, the lodge's first master and "prime mover," was an uncle of Fairfield's alleged UGRR operator, Daniel Mendenhall, who himself was a master mason in Fairfield by 1851. None of the Salem Freemasons is a known UGRR operator, but they held their first meetings in the hotel of Duvall W. Henderson, who used it as a UGRR station.[38]

---

[37] "Mr. Jefferson and the Ordinance of 1787," *Iowa Sentinel*, Sept. 18, 1847, p. 1, col. 6, reprinted from the *Albany (N.Y.) Atlas*. Lodge Master Dr. Myers later married a sister of Free-Soiler Mungo Ramsay, of Washington Co., Pa.

[38] Of the new Salem Lodge, No. 17, William Gordon was master, James Stephenson was senior warden, and Caleb Webster was junior warden. William Gordon's wife Deborah (Mendenhall) was a sister of Daniel Mendenhall's father Elijah. Daniel Mendenhall was a master mason in Clinton Lodge, No. 15 in 1851, and a junior deacon in 1852. Grand Lodge of Iowa Freemasons, *Proceedings of the Grand Lodge of Iowa: of the Most Ancient and Right Honorable Fraternity of Free and Accepted Masons, at its Several Grand Annual Communications*, Vol. I, Muscatine, Iowa: Raymond, Foster & Kystra, 1858, pp. 228, 243, 351, 432. Deborah (Mendenhall) Gordon died in 1885, and William M. Gordon married, second, Mary Calhoun, whose father William was a second cousin of Ripley, Ohio's UGRR station master Rev. John Rankin, and a first cousin of Birmingham, Iowa's UGRR operator Newton Calhoun. For Gordon as "prime mover" of Salem's Freemasons in D. W. Henderson's UGRR-station hotel, see *Fort Madison Evening Democrat*, June 20, 1949, p. 6, cols. 1-2.

Were the seven Fairfield Freemasons, or some of them, perhaps, Joseph A. McKemey's secret UGRR seven in 1847? Clearly more Fairfielders, some quite prominent citizens, were beginning to oppose slavery more firmly. However, at that time most Fairfielders were still anti-abolitionist and anti-UGRR, and in favor of continued conciliation of the South, if not themselves pro-slavery. A UGRR network would have had to be utterly secret and woven of bonds of absolute trust, likely involving close kin, old childhood friends, and possibly secret societies like the Freemasons.[39]

1847: Dred Scott and Fairfield, Iowa: Beginnings

Meanwhile, in a shocking setback in St. Louis, Missouri, the slave Dred Scott lost a lawsuit for freedom on June 30, 1847 on a technicality, and some peripheral actors in that drama—Almira Russell and Verplanck Van Antwerp—were rather closely tied to Fairfield, Iowa.

Dr. John Emerson, an army surgeon, had bought Dred Scott in St. Louis about 1832 and the following year brought him to Fort Armstrong on Rock Island, Illinois, and onto his land claim nearby in Bettendorf, Iowa, and from there in 1836 to Fort Snelling, Wisconsin Territory (now Minnesota). There Dred had met and married Harriet (Robinson), who had belonged to the fort's Indian agent, Lawrence Taliaferro. Both Dr. Emerson and Taliaferro attended the Chippewa land cession treaty of 1837 at Fort Snelling, and witnessed its signing, as did the treaty commissioner's secretary, Verplanck Van Antwerp.

The Scotts had returned to St. Louis in May 1840. Dr. Emerson left the army in 1842 and went to Davenport, Iowa, where he died after willing all his assets to his wife Irene (Sanford) in trust for their daughter. For several years, Irene hired out the Scotts as slaves to various employers in St. Louis, including wholesale grocer Samuel Russell. Dred tried to purchase his freedom from Irene for $300 in 1846, but she refused, and so Dred and Harriet went to court, for their own freedom and for their two children.

---

[39] Charles J. Fulton, *History of Jefferson County, Iowa: A Record of Settlement, Organization, Progress and Achievement*, Chicago: S. J. Clarke Publishing Co., 1912-1914, Vol. 1, p. 231. During "Bleeding Kansas" Missouri's pro-slavery "Blue Lodges" arose from Freemasonry; possibly Fairfield's anti-slavery UGRR had had Masonic roots as well.

By Missouri law, the Scotts' extended residence in the free state of Illinois and territory of Wisconsin had set them free, according to the "once free, always free" principle upheld in the Missouri courts from 1824 on. But when grocer Samuel Russell testified that he had hired the Scotts from Irene Emerson and paid her father—John F. A. Sanford of the American Fur Company—for their services, cross-examination revealed it had been Samuel's wife Adeline Russell who had hired the Scotts from Irene; Samuel had only paid Irene's father their wages. No one tried to overturn Missouri's time-honored "once free, always free" principle, but without a deposition from Adeline or proof that Samuel had known the Scotts were slaves, his testimony was considered hearsay, and the case was dismissed.

The Scotts' lawyers moved successfully for a retrial, but the case did not come to court until January 12, 1850. In the meantime Irene got the St. Louis County sheriff to take custody of the Scotts, hire them out, and collect their wages, while she moved to Massachusetts, where she met and eventually married the abolitionist Dr. Calvin C. Chaffee.

Samuel and Adeline Russell's teenaged daughter Almira, meanwhile, was a "frequent visitor to Fairfield ... at the home of Gen. V. P. Van Antwerp." Almira probably visited the Unionist Democrat Verplanck (or Ver Planck) Van Antwerp in Fairfield, Iowa before she married Lieutenant Winfield Scott Hancock in January 1850, as Verplanck was Fairfield's land office receiver from December 1845 only until May 1849, and had left Fairfield for Keokuk, Lee County, Iowa by 1850.[40]

Why was Almira Russell frequently coming from St. Louis to Fairfield before 1850? Perhaps she was visiting Verplanck's oldest daughter Catherine, who was close to her age; both were born around 1831. But Verplanck also must have known Dred and Harriet Scott personally. As he had been secretary to Treaty Commissioner Henry Dodge in Wisconsin Territory at the Chippewa Indian treaty of 1837, he had witnessed the treaty along with Dr. Emerson and Indian Agent Lawrence

---

[40] "Mrs. Almira Russell Hancock, widow of General Winfield S. Hancock, who died in New York last week and whose remains were interred at St. Louis, will be remembered by some of our older citizens as a frequent visitor to Fairfield years ago at the home of Gen. V. P. Van Antwerp." *The Fairfield Ledger*, April 26, 1893, p. 3, col. 6.

Taliaferro around the time "their" slaves Harriet and Dred Scott were getting married there.[41]

In the wake of the unjust finding of 1847 and her parents' inadvertent failure to help Dredd and Harriet Scott, had Almira come to Fairfield perhaps in part to attempt to obtain evidence from Verplanck of the Scotts' living in the free Wisconsin Territory, or perhaps even to effect the Scotts' freedom through the UGRR?

Again, Fairfield reportedly now had a secret UGRR ring consisting of Joseph A. McKemey and seven others. Fairfield openly had at least one Democratic friend of Blacks in George Acheson, the prosecuting attorney who had just refused to prosecute Charles Forrester for not posting a bond, and who would marry Joseph A. McKemey's sister-in-law that December. Almira may well have taken the Dred Scott case to heart; she had "that warm-hearted, unaffected way which makes the stranger feel at once that he is not a stranger."[42]

The Russell family held household slaves in St. Louis, but Almira DuBois Russell had some interesting anti-slavery ties. Her step-grandmother and namesake Almira (Pease) DuBois's first cousins Peter P. and Hiram Abif Pease founded Oberlin, Ohio and built Oberlin College, first in the country to admit both Black and female students. Peter Pease was a founding trustee of Oberlin along with the abolitionist Owen Brown, father of the famous John Brown. Almira Russell was born in Ohio. Her mother was from New York and her father was from Pennsylvania, like her husband, Winfield Scott Hancock: a Union Democrat who became one of the North's most famous generals in the Civil War.

Almira's friend Catherine Van Antwerp married George H. Williams, an anti-slavery Democrat who would become a Radical Republican Supreme Court judge, U.S. attorney general, and senator in Grant's administration, upholding Reconstruction and Black civil rights. Even though Verplanck Van Antwerp was land-office receiver in Fairfield under the Southern-sympathizing Register Bernhart Henn, he was a protégé of Martin Van Buren, the Free-Soil presidential candidate in 1848, and became a brevet brigadier general in the Union Army in the

---

[41] Catherine Van Antwerp and Almira Hancock were each 19 in the 1850 U. S. census: Catherine with her parents in Keokuk, and Almira both with her parents and with Winfield S. Hancock in St. Louis.

[42] *Fort Wayne Sentinel*, Feb. 19, 1886, p. 1, col. 3.

Civil War, serving on the staff of his old friend, General Winfield Scott Hancock.

Almira's parents both testified at Dred and Harriet Scott's retrial in 1850, corrected the technicality, and the Scotts won their freedom. But their erstwhile mistress appealed the decision, and for the next seven years the Scott case would be caught up in courts and become a symbol of the North and South's increasing acrimony over slavery.

1848: The Free-Soil Party Arises: Small Beginnings, Much Despised

The Free-Soil party, an amalgam of anti-slavery "Barn-burner" Democrats, anti-slavery "Conscience" Whigs, and the Liberty Party, arose in August 1848 in response to the Democrats' refusal to support the Wilmot Proviso, which would have banned slavery in any territory won in the Mexican War. Accordingly, the new party arose with the single issue of keeping slavery out of the territories. It nominated ex-President Martin Van Buren of New York for president and Charles Francis Adams, Sr., of Massachusetts, son of President John Quincy Adams, for vice president.

Jefferson County was still solidly Democratic and pro-Southern; in the presidential and vice-presidential elections of 1848, Democrats Lewis Cass of Michigan and William O. Butler of Kentucky got 739 votes, while Whigs Zachary Taylor and Millard Fillmore got 637 votes. Both Lewis Cass and Zachary Taylor were slave-holders. Cass favored popular sovereignty, allowing each territory to decide for itself whether it would be slave or free, a stance much favored by the pro-slavery faction. Taylor remained noncommittal, but favored preserving the Union at all costs. William O. Butler was a slave-holder, but opposed slavery's extension; Millard Fillmore held no slaves, but was a Southern sympathizer.

In Jefferson County, the Free-Soilers Martin Van Buren—an ex-slave-holder who had reversed his earlier pro-slavery policy as President—and Charles Francis Adams got only 21 votes, 1.5% of the total. Still a despised and negligible third party, the Free-Soilers would soon arise from obscurity, as the pro-slavery expansionist agenda revealed itself in the Fugitive Slave Act of 1850 and the Kansas-Nebraska Act of 1854.

1850: The Fugitive Slave Act

By conquering Mexico, the United States acquired vast new tracts of southern territory, as the pro-slavery contingent had desired. Now officially ceding Texas as well as New Mexico and Alta California, Old Mexico lost nearly half its territory to the United States.

However, California desired to enter the Union as a free state, though it lay partially south of the 36°30' latitude separating slave from free territory in the Missouri Compromise of 1820. As the South threatened again to secede, Henry Clay forged a new compromise permitting California's entry as a free state and the long-petitioned abolition of the slave trade in the District of Columbia for the North, balanced by a much stiffer Fugitive Slave Act for the South. The Act created and authorized state commissioners to remand Blacks to claimants producing "satisfactory proof" of ownership, and to summons any bystanders to help them capture Blacks, even in free states, and required all U. S. marshals and deputy marshals to enforce the act under penalty of a fine of $1,000—at least $23,000 in 2016—plus the value of the slave's lost labor. It also denied fugitives the right to give testimony at their hearings, and increased the penalty for anyone obstructing a claimant from arresting a fugitive, or aiding or concealing a fugitive, to a fine of $1,000 and six months in jail.

In short, the Act made every Northerner responsible for enforcing Southern slavery. Blacks who had escaped slavery and were living peaceably in the North now fled to Canada. Many freedmen now also left for Canada, as the Act also made it much easier for slave-catchers to kidnap free Blacks in the northern states and sell them into slavery in the South, in what was called the Reverse Underground Railroad.[43]

Shortly before the Civil War, Fairfielders would encounter what they believed to be a case of the Reverse Underground Railroad, and

---

[43] *"Great Excitement in Pittsburgh among the Blacks.*—The Pittsburgh Gazette of Tuesday says: The passage of the fugitive slave bill has greatly alarmed our colored population, a portion of which consists of escaped slaves, and many of the latter have left the city for Canada.... The ... law will have the effect of banishing the great majority of the escaped slaves to the British possessions...." *Burlington Weekly Telegraph*, Oct. 5, 1850, p. 3, col. 2. Free Blacks too also fled to Canada; there were more Free Blacks in 1860 than in 1850 only in Ohio, Michigan, and Illinois: all bordering on Canada, "where the runaway slave or the free man of color in danger could flee when threatened." Fred Landon, "The Negro Migration to Canada after the Passing of the Fugitive Slave Act," *The Journal of Negro History*, Vol. 5, No. 1 (Jan. 1920), pp. 22-36.

would take definitive steps to stop it. But that "kidnapping" drama occurred a full decade after the Fugitive Slave Act of 1850, and is best explored in chronological sequence. Much unfolded during that decade to shift many people's opinions more firmly against slavery.

Local Reactions to the Fugitive Slave Act of 1850

Anti-slavery Northerners perceived the Fugitive Slave Act, which they dubbed the "Bloodhound Law," as yet another example of the federal government's subservience to Southern aggression, and of its complicity in legalized immorality, and the Act stiffened the defiant yet secret resolve of many abolitionists on the UGRR.

That secrecy plus an almost four-year gap in extant Fairfield newspapers, from July 1849 to March 1853, hinders our assessment of how anti-slavery Fairfielders reacted at the time. The Whig (and later Republican) *Fairfield Ledger* had been established in November 1849 by Orlando McCraney, later run by A. R. Fulton and for nearly 50 years by W. W. Junkin, but their earliest thoughts in print on the Fugitive Slave Act are lost to us. They might not have had much written opposition for a time, as the Democratic *Sentinel* was suspended in 1851 upon the death of editor Robert B. Pope. It was resumed in 1853-54 by the fiercely pro-slavery David Sheward and in 1855-57 by him and his brother W. H. Sheward, but then folded, to be succeeded by the *Jeffersonian* in 1857-60, and David Sheward's *Constitution and Union* in 1861-64.

In nearby Mount Pleasant, abolitionist and UGRR station-master Samuel L. Howe lamented the Fugitive Slave Bill in his anti-slavery newspaper *Iowa True Democrat:*

> When this bill becomes a law, freedom is but a name. Ourselves, wives, and children, will be at the mercy of the slave-holder, without judge or jury.[44]

---

[44] Quoted in (Burlington) *Iowa State Gazette*, Sept. 25, 1850, p. 2, col. 2. Upon Howe's quote, the Democratic *Gazette*'s editor commented, "Better emigrate to Liberia, neighbor. You will be safe there." His response sounds eerily like the conservatives' rejoinder to the Vietnam War protesters: "America – love it or leave it." D. M. Kelsey established the *Iowa Freeman* in Mt. Pleasant in 1849. It was the only abolitionist newspaper then in the northwest. Samuel L. Howe became editor in 1850, and the newspaper was renamed the *Iowa True Democrat* by that April.

1853: Jefferson County's UGRR and Free-Soilers Grow Stronger

Despite and perhaps in defiance of the Fugitive Slave Act, Jefferson County's Congregationalists appear to have strengthened the UGRR lines through Jefferson County during this period. In May 1853, they established churches both in the UGRR hub of Salem and in Glasgow, which lies directly on the route from Salem to Fairfield, and is midway between them, nearly 12 miles from each. Mt. Pleasant's Rev. Simeon Waters organized the Glasgow Church; his wife had served on the UGRR back in Ohio, and he would soon be the Free-Soil candidate for Iowa's governor.[45]

The first minister at both the Salem and the Glasgow Congregational churches was the UGRR operator Rev. Joseph C. Cooper, a fervent member of UGRR patriarch Asa Turner's Denmark Church. Cooper's uncle, William B. Cooper, was also a UGRR operator in Denmark. Rev. Cooper presided over the "execution of King Alcohol" in Fairfield in January 1854 with Rev. Levin B. Dennis, the Methodist minister of the UGRR Byrkit family. Perhaps the two ministers cooperated in matters of anti-slavery as well as temperance, and it appears possible that Rev. Cooper may have helped fugitives from Salem through Glasgow to Fairfield, where the Byrkits would relay them on to Richland or Pleasant Plain.[46]

The Free-Soil Party ran candidates for Jefferson County's recorder, sheriff, and coroner in the election of August 1, 1853; all three lost spectacularly. Free-Soiler Benjamin C. Hoskins got 20 votes for recorder—5 from Walnut, 8 from Fairfield, and 7 from Round Prairie—while Whig Henry P. Warren beat Democrat Barnet Ristine, 698-676.

---

[45] Simeon's father-in-law Horace Ensign was a prominent UGRR director in Madison, Ohio, a founder of the Anti-Slavery Party and a close friend of the nation's foremost abolitionists. He helped found the Congregational Church in Madison, and helped rescue Milton Clark from slave-hunters in 1842: *Ashtabula, Geauga and Lake...*, Chicago: Lewis Pub. Co., 1893, p. 898. Simeon's wife, Frances Almira (Ensign), had also served on the UGRR: "Especially was she the 'friend of the friendless.' No appeal from any suffering or oppressed human being ever failed to awake her warmest sympathy. The fugitive from slavery, the freeman of color in her native State, Ohio reaped especially the benefit of her unwearied energies and her warmest prayers." Obituary in *The National Era*, July 22, 1852, p. 111. For Simeon Waters as founder of Glasgow's Congregational Church on May 21, 1853, see *Fairfield Ledger*, July 13, 1876, p. 3, col. 5; "Simeon" is there misspelled "Simon."

[46] William Henry Perrin, ed., "William Brewster Cooper," *History of Effingham County, Illinois*, Chicago: O. L. Baskin & Co., 1883, part 2, p. 11.

Free-Soiler John Ely got 7 votes for sheriff—all from Round Prairie—compared to 703 for Whig G. M. Chilcott and 596 for James A. Galliher. Free-Soiler Jonathan Hoskins, Sr. got no votes for coroner—apparently not even voting for himself—while Democrat John Steel beat Whig A. H. Brown, 645-585. Surprisingly, no one in the Quaker stronghold of Penn Township voted Free-Soil, though they always favored the Whig candidate over the Democrat. In Jefferson County's other two races, Democrat David C. Cloud narrowly beat Whig Samuel A. Rice for attorney general, 529-526, and Democrat Sam Jacobs beat Whig W. Z. Hobson for surveyor, 730-566.[47]

But perhaps partly in response to the Fugitive Slave Act, by that 1853 election the county's Whig Party had become more overtly Free-Soil; G. M. Chilcott, Henry P. Warren, Asahel H. Brown, and William Z. Hobson all soon became active anti-slavery Republicans. The Free-Soil tenor of the new nominees was not lost on the Democratic *Sentinel.* As *Ledger* editor A. R. Fulton commented drily:

> The Sentinel manifests evident symptoms of being displeased with the ticket formed by the Whig convention on Saturday week. Well, never mind, the Whigs no doubt would have been glad to have pleased the Sentinel, if by doing so they could have done justice to themselves. The Sentinel has nothing to say as to the merits of the men that form our ticket, but it is of the opinion that a portion of the Whig party is sold, for the purpose of securing the Freesoil vote.

A. R. Fulton did not deny the candidates were Free-Soil, only that the ticket was made unfairly, or with dishonest or incapable candidates:

> Neither the Whig party nor any portion of it is sold, and this clamoring of the Sentinel is all moon-shine. There never was a ticket made more fairly than that of the recent whig convention. Every man on the Whig ticket is known to be upright, honest and capable. Perhaps the Sentinel would prefer them otherwise. But we cannot help it now, and we would not if we could.[48]

Fulton's defiant resistance to what he perceived as the corrupt "Slave Power" was now palpable. He and W. W. Junkin would do consummate service in awakening Fairfielders to the evils of slavery in

---

[47] *Fairfield Ledger*, Aug. 4, 1853, p. 2, col. 6.
[48] *Fairfield Ledger*, July 14, 1853, p. 2, col.

print; whether they also resisted physically, perhaps on the UGRR, remains unknown.

Other ardent abolitionists who came to Jefferson County around 1853-54 were Jane (Dillingham) and Richard Gaines, who had reportedly served on the UGRR in Cincinnati and probably continued their work here in Black Hawk Township. Jane's parents were noted Quaker UGRR station masters in Ohio, and even before the Fugitive Slave Act had passed, Jane's brother Richard Dillingham had been arrested in Tennessee for "Negro Stealing" in 1848, and had tragically died in prison of cholera in 1850: a martyr to the cause, eulogized by John Greenleaf Whittier and Harriet Beecher Stowe.[49]

To avoid the horrific consequences of the Fugitive Slave Act, one could perhaps still plead ignorance when assisting a fugitive slave. On November 10, 1853, William H. Bradney of Jefferson County's Des Moines Township, west of Libertyville, posted a notice in the *Fairfield Ledger*:

> To the Public. On the Twentieth of October last, I purchased a MARE from a negro, who stated he was from Missouri. It has since been supposed that he was a runaway slave, and the mare a stolen animal. If such be the case, of course the owner has a better right to the mare than myself, and by proving his property can have her. In such event I will claim expenses. Residence in Des Moines Township, Jefferson co., Iowa. Wm. H. BRADNEY.[50]

If the mare's seller was indeed a fugitive slave, as he probably was, the notice implies he was trying to escape on his own, without benefit of the Underground Railroad. But William Bradney waited three weeks

---

[49] Obituary of Richard Gaines in the Pueblo, Colorado *Weekly Chieftain*, Sept. 11, 1879, http://www.kmitch.com/Pueblo/obits/gaa-galk.html. For Richard Dillingham's UGRR activity and trial, see Wilbur H. Siebert, *The Underground Railroad from Slavery to Freedom*, New York: The MacMillan Co., 1898, p. 174 ff.; A. L. Benedict, *Memoir of Richard Dillingham [of Morrow Co., Ohio], With an Introductory Poem by J. G. Whittier*, Philadelphia: Merrihew & Thompson, 1852, in Wilbur H. Siebert UGRR Collection, Ohio History Society, https://tinyurl.com/ohiosiebert. Richard Dillingham's letters to Jane and Richard Gaines are on pp. 23-25 [original pp. 14-15], and his attorney's post-mortem letter to Richard Gaines is on pp. 36-37 [original pp. 22-23]. Much of his correspondence is also in Harriet Beecher Stowe, *A Key to Uncle Tom's Cabin; Presenting the Original Facts and Documents upon which the Story is Founded, together with Corroborative Statements Verifying the Truth of the Work*, Boston: John P. Jewett & Co., 1853, pp. 55-59.

[50] Ad dated Nov. 10, in *Fairfield Ledger*, Nov. 24, 1853, p. 4, col. 5.

before attempting to notify the mare's owner, suggesting he may have wished to give the owner's other "property" ample time to escape to Canada. The Bradneys had recently arrived in Iowa from Adams County, Ohio, an area teeming with UGRR activity. William may have claimed not to have known the Black Missourian was a fugitive, in order to avoid the Fugitive Slave Act's heavy punishment of anyone in any state who did not actively help Southern slave-catchers recover "their" property. The Fugitive Slave Act began to awaken some Iowans to the aggressive nature of Southern slavery, but more would be galvanized by the Kansas-Nebraska Act of 1854.

1854: The Kansas-Nebraska Act

The United States wanted railroads to the west coast, and the railroad companies needed settlers in the vast lands the tracks would have to cross. When the U. S. House of Representatives passed a bill to organize the "Nebraska Territory"—which then included modern-day Kansas, Nebraska, and parts of Colorado, Idaho, North and South Dakota, Wyoming, and Montana—in early 1853, the fire-eating Missouri Senator David Rice Atchison refused to support the bill unless the territory allowed slavery, which the Missouri Compromise did not permit. All the Southern senators followed Atchison's lead, tabling the bill. Democratic Senator Stephen A. Douglas from Illinois introduced a compromise in January 1854 which split the territory into Kansas and Nebraska, with the slavery issue in each territory to be decided by popular sovereignty. The Senate passed the Kansas-Nebraska bill on March 4 by a vote of 37-14, supported by both of Iowa's Democratic senators, Augustus C. Dodge of Burlington and George W. Jones of Dubuque: Both had Missouri roots, and Jones had held slaves even in Iowa. And after much debate and some physical violence, the House passed it in May, by 113 to 100, supported by Iowa's Democratic Representative, Bernhart Henn of Fairfield.

Dubbed the "Nebraska Outrage" by Free-Soilers, the Kansas-Nebraska Act of 1854 repealed the 1820 Missouri Compromise forbidding slavery north of the 36°30' parallel in the Louisiana Purchase territory except for Missouri. Now the settlers themselves would vote on whether their new territories would be slave or free.

As mentioned, there was a precedent for decision by popular sovereignty; California lay partly south of the parallel, but had upon request entered the Union as a free state in 1850, together with the stiffer Fugitive Slave Act to appease the South. And Utah and New Mexico had also both been admitted as territories then, with the slave-vs.-free issue to be decided by popular sovereignty.

But Kansas and Nebraska were both part of the original Louisiana Purchase, both above the parallel, and thus were already free. They were also in Missouri and Iowa's backyard, and while Missouri fiercely desired slavery on her western border to reduce escapes and expand her slave-market, Iowa felt bullied and betrayed by the repeal, and just as fiercely resisted the prospect of more slave-state neighbors and more UGRR routes crisscrossing the state. If both Kansas and the immense Nebraska Territory admitted slaves, then the Northeast would essentially become an island of free states surrounded by an ocean of slavery, both geographically and politically overpowered by the Slave Power.

1854: *Whig and Republican Repercussions*

A statewide convention of Free-Soilers met on January 14, 1854 at Washington, Iowa and nominated Rev. Simeon Waters of Henry County for governor. The Free-Soilers were still a minor and despised third party; Waters had no hope of winning, but they had made a stand for freedom. As mentioned, Simeon had strong UGRR ties in Ohio, and on May 21, 1853 had organized the Congregational Church in Glasgow, Round Prairie Township, with UGRR operator Rev. Joseph C. Cooper as its first pastor.[51]

The Whigs of Jefferson County met in Fairfield on February 18, 1854, appointed Col. Isaiah W. McManaman president and Robert F. Ratcliff secretary, and elected delegates to attend the Whig State Convention in Iowa City. The delegates were Dr. Wesley Johnson Green, Col. John Park, James H. Hendricks, L. T. Gillet, Thomas Moorman, James M. Strong, Abner Frazier, Thomas Wamsley, William S. Lynch,

---

[51] Simeon's father-in-law Horace Ensign was a prominent UGRR director in Madison, Ohio, a founder of the Anti-Slavery Party and a close friend of the abolitionists Theodore D. Weld, Joshua R. Giddings, Benjamin F. Wade, William Lloyd Garrison, and James G. Birney. He helped found the Congregational Church in Madison, and helped rescue Milton Clark from slave-hunters in 1842. For Simeon Waters as founder of Glasgow's Congregational Church on May 21, 1853, see *Fairfield Ledger*, July 13, 1876, p. 3, col. 5; "Simeon" is there misspelled "Simon."

Alexander R. Fulton, Caleb Baldwin, Samuel Robb, J. S. Stuff, and Henry B. Mitchell, almost all of whom became prominent Republicans.

Iowa's Whigs met on February 22 in Iowa City and nominated the anti-slavery James W. Grimes for governor, and the Free-Soilers' Rev. Simeon Waters for secretary of state. Grimes soon came to Denmark, Iowa and courted the Free-Soilers. Historians generally consider an anti-slavery Whig meeting in Wisconsin on March 20, 1854 as the founding of the Republican Party, but Iowa's anti-slavery sentiment was already building into what would become some of the Republican Party's most significant early victories. On March 28, a coalition of anti-slavery delegates met in Crawfordsville at the Old Seceder Church and in the hotel and UGRR station called the "House of All Nations." The delegates included UGRR station master Samuel L. Howe and his son Edward from Mount Pleasant, UGRR operator Joel C. Garretson and Eli Jessup from Salem, and alleged UGRR operator Dr. Curtis Shedd from Denmark. At that meeting, Denmark's venerable UGRR-founder Rev. Asa Turner persuaded them to withdraw their nomination of Waters and support Grimes for governor.

James W. Grimes began canvassing Iowa to clarify his positions, after hearing he had been branded a "falsifier" by Iowa's Democratic Senators August Caesar Dodge and George Wallace Jones with Georgia's Senators Robert Toombs and W. C. Dawson. With prior announcement he came to Fairfield on May 31 and spoke at the courthouse to "a large number" of Jefferson County citizens, many of whom "were in from the country" despite the busy season. Grimes recommended amending Iowa's constitution to let the people elect the Supreme Court judges, to permit bank charters, and to hold state elections on a more convenient date. He spoke strongly against the repeal of the Missouri Compromise, and "denounced with just severity the cringing spirit of northern DOUGH-FACEDNESS in Congress, which yields a co-operation and willing submission to every measure, however unjust, for the special benefit of the South."

Grimes warned that Nebraska would remain slave terrritory, and opined that were it not for the Missouri Compromise, Iowa might now be a slave state, and that "slavery propagandists" would next assume "the right to establish slavery in the FREE STATES—basing their arguments on the ground that men have a right to carry their property into any State they please, and that as slaves are property, the owners or masters have a right to carry them into the Free States and hold them there." Northern men who voted to repeal the Missouri Compromise "would be found ready and willing to ... vote for any measure to enhance the slave interest." Grimes closed by "vindicating himself against the attacks" of Dodge and Jones. His speech was "a masterly exposition of principle and a triumphant vindication of the great measures of reform advocated by the Whig party of this State," and Junkin reported that Grimes's audience "expressed themselves highly entertained." All were impressed, except for "our democratic friend, 'Moses'"—perhaps Moses C. Shamp.[52]

At the state Whig Congressional Convention in June 1854, Jefferson County's delegates were J. F. Wilson, Caleb Baldwin, E. C. Hampson, James Beatty, and T. D. Evans, who was also on a three-person committee examining credentials and reporting the delegates in attendance, and was also appointed vice president of the convention. Caleb Baldwin was one of a three-person committee to report permanent officers of the convention, and James F. Wilson was Jefferson County's member of the committee to draft resolutions. The convention resolved:

1. Though favoring speedy organization of the Kansas and Nebraska territories, they regarded the just-passed Bill repealing the Missouri Compromise and opening the territories to slavery as violating faith and betraying the interests of freedom and the North, "tending ... to involve the country in commotion and strife," and declared that the people of that district would not endorse their congressmen in voting for that provision;

2. They opposed Senator Douglas's proposal to lay tonnage duties on Western river and lake commerce, but favored appropriations by the general Government;

3. They favored donating public lands to actual settlers, to keep out speculators "and advance the interests of the West;"

4. They highly condemned the Democratic Congress for refusing to pass the Homestead Bill to give every actual settler a quarter section of land;

5. They welcomed "to our broad prairies and free homes all of the oppressed of every nation and clime," and claimed all of the rights and privileges of citizenship for the foreigner declaring intent to become a citizen, "Senators Dodge, Jones and Butler, to the contrary notwithstanding;"

6. They denied and unqualifiedly disapproved of a Southern Democratic Senator's recent declaration in the Senate that Iowans would pre-

---

[52] *Fairfield Ledger*, June 8, 1854, p. 2, cols. 1-2.

fer a population of slaveholders with their slaves to a population of Germans and other emigrants from Continental Europe;

7. That Senators Dodge and Jones, by receiving over $1,000 each from the public treasury for travel mileage never performed, had shown themselves "unworthy of the confidence" of Iowans;

8. They had "entire confidence" in the ability and integrity of James W. Grimes, their candidate for governor, and pledged him their support. After seven ballots, R. L. B. Clarke emerged as their candidate for Congress, and Caleb Baldwin was appointed to the central committee of the district.[53]

UGRR operator Richard Gaines and other free-soil Democrats met in a Free Democratic Convention in Fairfield on June 24, 1854, and also resolved to support James W. Grimes for governor; most of these members later joined the nascent Republican Party. The abolitionist Rev. Charles H. Gates of Fairfield's Congregational Church spoke at the convention on the moral bearings of slavery, and was accused by the Democratic *Sentinel* of "treason against the constitutional institutions of our country in seeking to unite Church and State," and also of being a wolf in sheep's clothing.[54]

The mainline Democratic Party had apparently surrendered utterly to the wealthy and powerful "Slavocracy" and its urge for infinite expansion. A. R. Fulton and W. W. Junkin reported in the *Ledger* that prominent Democrats were indeed pushing to allow slaveholders to keep their slaves in Iowa, just as Iowa's Freesoiler Whig Gubernatorial candidate James W. Grimes had predicted in Fairfield:

> When Mr. Grimes asserted in his Speech delivered in the Court House in Fairfield, that the next step ... by the Democratic leaders and their southern confederates, would be the advocacy of the right of

---

[53] *Fairfield Ledger*, June 15, 1854, p. 3, cols. 1-2.
[54] "Iowa Free Soil Ticket," *Burlington Weekly Telegraph*, Feb. 4, 1854, p. 1, col. 7. O. A. Garretson claimed that the Liberty Party, Free-Soil Party, and others against slavery met in Crawfordsville in Feb. 1854 and birthed the Republican Party at least a month before the party's official birth at Ripon, Wisconsin on March 20, 1854: O. A. Garretson, "Travelling on the Underground Railroad in Iowa," *Iowa Journal of History and Politics*, Vol. 22, No. 3 (July 1924), pp. 452-453. Contemporary newspapers failed to mention such a convention, and it appears Garretson misdated the Crawfordsville convention of March 28, 1854. See Emory H. English, "Iowa Republicans Organized in 1856," *Annals of Iowa*, vol. 32, no. 1 (Summer 1953), pp. 43-46, http://ir.uiowa.edu/annals-of-iowa/vol32/iss1/5/. *Fairfield Ledger*, July 6, 1854, p. 2, cols. 2-4.

slaveholders to bring their slaves into free States and hold them, the wise men of the Democratic party pretended to brand him with stating that which he did not believe. How rapidly have they changed ... for now some of those very men, are advocating the right of slaveholders to hold slaves in all free States. In the town of Fairfield that doctrine has been advanced by prominent members of the Democratic party.... Let them alone and the fifteen years which Stephens of Georgia, said would make a slave state of Iowa, will but commence ... before the prediction will be fulfilled.[55]

"The doctrine," the editors continued, "has been advocated in the halls of Congress time and again by Southern members; it has now been caught up by the administration leaders of the North, and soon will become the settled policy of the Democratic party" and a plank in the Democratic platform, to be claimed by the likes of Iowa's Missouri-born Democratic Senator A. C. Dodge, who had first introduced the Nebraska Bill in 1853, as the doctrine of our founding fathers:

> Not content with having cursed our country with slavery to the present alarming extent, they will seek to make every territory a slave territory---every State a slave State. Slavery is to be nationalized if the present Democratic leaders, north and south, can retain the power. The curse is to be spread over the entire Union ... unless the people at once check the rampant spirit of slavery propagandism which has seized upon the leaders of the Democratic party.

Messrs. Fulton and Junkin called upon their readers to "take warning in time" to check the evil's spread:

> Think not what we stated is false; for we tell you solemnly it is true.... When you come to deposit your vote in the ballot-box on Monday next, think of the doctrine that slave-holders have a right to bring their slaves into Iowa and hold them -- that Iowa shall thus be made a slave State.

The editors then pointed out what that would mean for Iowans:

> Reflect that ... that doctrine would bring the slave labor of the south in competition with the free labor of our own beloved State. That the labor of yourselves, your fathers and your sons, would be forced into

---

[55] *Fairfield Ledger*, August 3, 1854, p. 2, col. 5.

> a competition with the labor of the degraded slave. That Iowa shall become a slave-breeding State to supply the demand for slaves on southern plantations!

This misery, obviously, was not to be tolerated in Iowa:

> Bah! such a doctrine to be advocated by free men in a free State! Put your curse upon it. Sink it below the reach of resurrection; stamp it with the infamy it deserves.

In the previous column, the editors had lauded the coalition of Free-Soilers, Whigs, and Free Democrats to oppose slavery's expansion, and denied the *Sentinel*'s "unmitigated falsehood" that R. F. Ratcliff, J. F. Wilson, and A. R. Fulton had "sold out" the Whig Party to the despised abolitionists at an "abolition convention" at Pleasant Plain: The Whigs would not be seeking to abolish slavery, only to curtail its expansion. Fulton and Junkin conclude with a stirring call to action at the polls:

> Let these fool-hardy men know that such a doctrine can meet with no toleration in free Iowa. Tell them at the ballot-box that you will now take the leadership in your own hands, and give them to understand that they must act as freemen or give up all idea of place and power.[56]

Now, more Iowans listened and agreed; truly the Kansas-Nebraska Act appeared to expose the South's plan to permit slavery to expand *ad infinitum*, even into the Northern states. In August 1854, James W. Grimes was elected governor, decisively ending the pro-slavery Democrats' 15-year rule in Iowa. Jefferson County, which had separated from the more Whiggish Henry County back in 1839 through its pioneering Democratic legislator, Col. William G. Coop, now voted the straight Whig ticket, electing Edmund Mechem, James Wamsley, and Robert Stephenson as state representatives, Robert F. Ratcliff as clerk, Caleb Baldwin as prosecuting attorney, and A. R. Fulton as surveyor. The pendulum continued to swing against the old-guard Democrats; within a few months Congressman Henn retired, and Senator A. C. Dodge lost to Free-Soiler James Harlan of Mount Pleasant.[57]

---

[56] *Fairfield Ledger*, August 3, 1854, p. 2, cols. 4-5.

[57] However, at the Iowa State Democratic convention in Iowa City on Jan. 8, 1856, B. Henn was chosen a delegate to the national Democratic convention: *Fairfield Ledger*, Jan. 17, 1856, p. 2, col. 7.

The victories were the triumphant swan-song of the Whigs, accomplished only through their coalition with the anti-slavery Free-Soilers and Free Democrats: in large part through the tireless efforts of Rev. Asa Turner, Denmark's UGRR patriarch. Though elected as a Whig, Governor Grimes helped found the nation's Republican Party over the next two years. However, Iowa still had plenty of pro-slavery conservatives, and UGRR activity in Iowa was still illegal and dangerous. Secrecy was essential.

Underground Railroad Codes

While there is no contemporary evidence to support the 1980s theory that UGRR operators used quilt-patterns as a code, an operator would sometimes pass a coded letter to alert the next station master that fugitives were on their way, specifying how many and sometimes what sex and complexion they were. G. W. Weston in Low Moor, Iowa wrote to C. B. Campbell in Clinton, Iowa:

> Mr. C. B. C.   Low Moor, May 6, 1859
> Dear Sir – By tomorrow evening's mail you will receive two volumes of the 'Irrepressible Conflict' bound in black. After perusal please forward and oblige,
> Yours truly, G. W. W.

Wilmington, Delaware's Quaker station master Thomas Garrett would usually write, "I send thee two, three or more bags of black wool."

Ex-Fairfielders H. M. "Hub" Hoxie and Thomas Mitchell in Mitchellville, Iowa wrote to J. B. Grinnell in Grinnell, Iowa:

> *Dear Grinnell:* -- Uncle Tom says if the roads are not too bad you can look for those fleeces of wool by to-morrow. Send them on to test the market and price, no back charges.
> Yours, HUB.[58]

---

[58] All three coded letters are in Wilbur H. Siebert, "The Underground Railroad," *New England Magazine*, New Series Vol. 27, No. 5 (Jan. 1903), p. 572; further details in James Patrick Morgan, *John Todd and the Underground Railroad: Biography of an Iowa Abolitionist*, Jefferson, North Carolina: McFarland & Co., 2006, p. 86.

1854-1855: *Sheep, Wool, and Abolitionism in Fairfield*

Several of these coded UGRR letters mention "wool." Northern sheep-raisers and wool-producers tended to be early abolitionists, and many famous abolitionists and UGRR managers like Levi Coffin, Thomas and Charity Rotch, John Brown, and Josiah B. Grinnell favored sheep-raising and wool-manufacture as an alternative to slave-grown cotton.

At the first Iowa State Fair, held in Fairfield in October 1854, the "original abolitionist" Quaker John Andrews of Penn Township, and the UGRR operator Henry Morgan of Clay, Iowa both won prizes for their sheep. Benjamin C. Perkins brought his pen ewes and Saxony sheep all the way from Lorain County, Ohio to win a diploma and a commendation. His second cousin Emily Perkins had married the abolitionist Edward Everett Hale, who had co-founded the Massachusetts Emigrant Aid Society in March 1854 to bring immigrants from New England to make Kansas a free state. Benjamin's third cousin Simon Perkins was a partner in the wool business with the fiery abolitionist John Brown, and all three men had judged sheep together at Ohio State Fairs.[59]

As Border Ruffians in Missouri began harassing and repelling Free-State settlers attempting to reach Kansas through their state, anti-slavery emigrants increasingly began to travel to Kansas through Iowa instead. Perhaps B. C. Perkins was also judging Fairfield's merits as a supply-depot for Free-Soil emigrants on behalf of his colleague, John Brown, who would himself soon pass through Iowa repeatedly on his way to and from Kansas. John Brown was interested both in a Free Kansas and in UGRR routes through Iowa. his second cousin Manning Mills was a UGRR operator in Clay, like Henry Morgan, and Manning's brother Oliver Mills moved from the UGRR nexus of Denmark, Iowa to serve the UGRR in Lewis, Iowa, like ex-Fairfielder George B. Hitchcock.

---

[59] B. S. McElhinny and J. M. Shaffer, "History and Proceedings of the First Fair of the Iowa State Agricultural Society, held at Fairfield, October 25, 1854" in John R. Shaffer, *Report of the Iowa State Agricultural Society for the Year 1874*, Des Moines: R. P. Clarkson, 1875, pp. 523-524. Perkins's name was given as "P. C. Perkins" of "Loraine" Co., but this was doubtless an error for the well-known sheep-farmer B. C. Perkins. We have found no P. C. Perkins involved in sheep-raising in Lorain County at that period.

In Fairfield, too, not only sheep-raising, but wool-manufacturing also had strong anti-slavery overtones. Fairfield's wool-carder Elisha Wetmore had anti-slavery and UGRR ties in Salem, Iowa, where his nephew Nelson Gibbs and brother-in-law Reuben Dorland both helped Ruel Daggs' slaves escape recapture in 1848. Fairfield's abolitionist Congregational Reverend Charles Gates was the son of a Massachusetts woolen manufacturer with ties to other prominent abolitionists. And Fairfield's coverlet-weaver Daniel Stephenson apprenticed with and worked for UGRR and Kansas Free-Soiler families. In 1864 he named his son *Wilson Grant* after both civil-rights heroes, Fairfield's James F. Wilson and Ohio's Ulysses S. Grant.[60]

1855-1856: Ulysses S. Grant, Sheep, and Fairfield

Though the Civil War officially started in 1861, it unofficially began with the Kansas-Nebraska Act of 1854, when Americans first began killing each other over slavery. Several of Ulysses S. Grant's contemporaries reported that he came to Fairfield, Iowa for about nine months in 1855-56, with a large herd of sheep which he was taking west. He had resigned as captain from the Army and returned from California to his in-laws' plantation in Missouri in 1854, just as the Kansas-Nebraska Act was passed.

Fairfielders may well have later confused Ulysses with another "Captain Grant"—California steamboat Captain James A. Grant, who did visit Fairfield in June 1855. That Captain Grant's activities in Iowa remain obscure. But Ulysses' father had a deep, lifelong friendship with the sheep-expert and UGRR manager John Brown, who was then crisscrossing through Iowa to and from Kansas. Also, Ulysses probably served on the UGRR as a boy, and was certainly experienced at moving men and materiel as an Army quartermaster. He had connections in Jefferson County to those sharing his father's abolitionist stance, and a surprising number of people in Fairfield had UGRR connections to Ulysses' hometowns in Ohio.

Given all that, it also seems possible that Ulysses may indeed have come to Iowa and secretly helped supply Free-Soilers en route to Kansas. If so, he performed an unsung service for freedom even in the first

---

[60] In the Civil War, Daniel's brother Abel enlisted in Co. E, 2nd Iowa Infantry, but was killed at the Battle of Corinth, Mississippi in 1862.

beginnings of the Civil War. That story is complex and better saved for another book; this one notes some of the more peripheral connections in Fairfield and Jefferson County, Iowa to U. S. Grant.[61]

March 1855: Democratic Election Fraud in "Bleeding Kansas"

The Kansas-Nebraska Act polarized nearly everyone, and birthed both the North's Free-soil Republican Party and the South's pro-slavery Masonic-based "Blue Lodges" and the Knights of the Golden Circle: probably the predecessors of the Ku Klux Klan, as *Circle* in Greek is *Kuklos*. As the slavery question was left to the settlers themselves, the Midwest and New England sent companies of Free-Soil settlers, while Missouri Senator David R. Atchison and his Southern allies sent some pro-slavery settlers and thousands of violent Border Ruffians who were not settlers, but who poured into Kansas and threatened election judges, intimidated legitimate voters, and stuffed ballot boxes at gunpoint on March 30, 1855. The town of Leavenworth recorded five times as many votes as its entire population, with 90% of the votes being pro-slavery, though a majority of Kansas residents were anti-slavery.[62]

The news of the election fraud soon reached Fairfield, prompting W. W. Junkin to write:

> In our paper today [is] news concerning the recent elections in Kansas. Dark and gloomy are the prospects of Freedom in that Territory. Swarms of renegade Missourians have again trampled underfoot the sacredness of the ballot box, and forced upon the actual settlers of Kansas a pro-slavery legislation. There is nothing now in the way of establishing slavery on the soil which the Missouri Compromise made free, and we may now look for the speedy erection of the black standard of human bondage, where for thirty years the tree of liberty was striking deep roots to withstand the storms and tempests of the future.

---

[61] For Ulysses S. Grant on the UGRR, see G. L. Corum, *Ulysses Underground: The Unexplored Roots of U. S. Grant and the Underground Railroad*, West Union, Ohio: Riveting History, 2015.

[62] The Free-Soilers condemned the newly-elected "Bogus Legislature" and established their own government at Topeka. President Franklin Pierce branded them as treasonous, and Free-State Governor Charles Robinson was arrested and imprisoned for four months in 1856 before being acquitted a year later. Jason Roe, "The Contested Election of 1855," Kansas City Public Library, *Civil War on the Western Border: The Missouri-Kansas Conflict 1854-1865*, https://tinyurl.com/election1855.

Junkin reminds us of the politicians who lied in saying that "popular sovereignty" would inevitably result in Kansas becoming a Free State:

> While the Nebraska-Kansas bill was pending in Congress, we were told that its passage would not affect the freedom of Kansas—that Kansas would remain free forever. What now do we see? Is not the desperate slave spirit of the despicable Atchison and his swarms of perjured minions, who reside in Missouri and vote in Kansas, running riot over the fair and fertile plains of that territory, carrying the blight and curse of slavery wherever they go? Is it thus that Kansas is to be made free? Is this the safeguard that the "squatter sovereignty" throws around the dearest interests of equality and freedom?

Junkin drives home the relentless aggression of the Slave Power:

> Slavery prompted the repeal of the Missouri restriction; slavery procured the passage of the bill repealing the restriction; and slavery is now availing herself of that repeal to extend her terrors where they never could otherwise have gone. Kansas will be a slave State; and it will avail those who are responsible for the act nothing to attempt to disguise the fact longer. The historian may already write the first chapter of the history of slavery in Kansas.[63]

And indeed, as Junkin predicted, on September 15, 1855 the pro-slavery "Bogus Legislature" in Lecompton would enact draconian laws making any sort of UGRR activity punishable by death. Resisting an officer attempting to arrest a slave was punishable by at least two years at hard labor. Publishing, printing, or even bringing into Kansas a book or pamphlet calculated to produce "rebellious disaffection" among slaves, or induce them to escape or resist authority, was punishable by at least five years at hard labor. Writing or saying that persons have no right to hold slaves in this territory was punishable by at least two years at hard labor. The laws were all enforced by President Pierce's complaisant Southern-sympathizing administration.[64]

---

[63] *Fairfield Ledger*, April 12, 1855, p. 2, col. 1.
[64] The Lecompton legislature's pro-slavery laws are in "Border Ruffian Laws in Kansas," *Subduing Freedom in Kansas: Report of the Congressional Committee, Presented in the House of Representatives on Tuesday, July 1, 1856*, New York: Greeley & McGrath, Tribune Office, pp. 28-31.

Junkin does not hesitate to place the responsibility for the Kansas debacle firmly where it belonged: with the pro-Southern Democrats:

> For this the Democratic party is responsible. That party opened the door and invited slavery to take possession of the domain of Freedom. Let not the leaders of that party shrink from the responsibility which they assumed when they sold out to the slavery propagandists of the South.—Let them stand up to the work which they agreed to perform. Let them stand by the terms of their contract. Let them do all this that those who are the true friends of freedom may no longer be lulled into a false security by supposing that all northern men are opposed to the extension of slavery.

And finally, Junkin points out that same complacency in Jefferson County had just permitted the Democrats to win the state elections, and he reaffirms his faith that Fairfield voters acted in ignorance, and were actually anti-slavery:

> When the people of Jefferson county, at the recent election, endorsed the Democratic State ticket, did they think that that endorsement carried along with it an approval of the outrageous effects which are now flowing from that great fountain of political and moral turpitude—the Nebraska bill?—We do not believe that the people of this county did take this fact into consideration, but we are satisfied at the same time that their act will be construed in that manner, however much they may protest their innocence. We know that such a construction would place the people of this county in a false position: for we believe that they are thoroughly anti-slavery in their sentiments. This shows the necessity of being ever on the guard, and if the people do not wish to be ranked with those who endorse the effects of the Nebraska bill now developing themselves in Kansas, they must be careful how they handle "regular Democratic tickets."[65]

June 2, 1855: Anti-slavery Democratic Convention in Fairfield

On May 10, 1855, a Vigilance Committee of Free-Soilers called for a convention on June 2, 1855 at the Fairfield courthouse to nominate candidates for the August election. The three Vigilance Committee

---

[65] *Fairfield Ledger*, April 12, 1855, p. 2, col. 1.

members signing the notice were John McCleery, Samuel Henderson, and Mungo Ramsay.[66]

On June 2, 1855, the Free-Soilers' "Independent Democratic" Convention met in Jefferson County with UGRR operator Richard Gaines as president and John McCleery as secretary. They resolved that they believed the Constitution to be an anti-slavery document, which "had it been faithfully administered, slavery would have long since ceased from our country;" that slavery was a state institution; that Congress had no power to pass a Fugitive Slave law; that the Federal courts enforcing that law represented a dangerous centralization of power; and that they would nominate only "true men [who had] always been right upon... human rights" and would do their duties with "credit to themselves and honor to their constituency." They nominated Ira G. Rhodes for county judge, John Williams for sheriff, Richard Gaines for treasurer and recorder, John McCleery for coroner, and R. Gaines, John Ely, and John McCleery for "vigilance for the ensuing year"—presumably against election fraud like that perpetrated in Kansas.[67]

June 23-26, 1855: A Freedom Seeker Captured in Burlington

On Saturday, June 23, 1855, while Ulysses S. Grant's first cousin Silas A. Hudson was the Whig mayor of Burlington, Missouri slave-catchers wielding revolvers and Bowie knives captured a freedom seeker and his UGRR conductor, Dr. Edwin James, as they were taking the ferry from Burlington across the Mississippi to Illinois. Naming the fugitive "Dick," and saying that he belonged to one Mr. Rutherford in Clarke County, they forced the pair back to Burlington, where "Dick" was thrown in prison and word sent to Rutherford. At the trial next Tuesday, Mayor Hudson had guarded the courthouse door from the press of the crowd, many of whom were proslavery sympathizers. But the first witness was Rutherford's son, who surprisingly swore that the accused was not the slave named Dick who had fled from his father. A great shout went up and a thousand men accompanied "Dick" and his conductor to the river to speed him on to Canada.

Governor James W. Grimes of Burlington wrote,

---

[66] *Fairfield Ledger*, May 31, 1855, p. 2, col. 2: notice dated May 10, 1855.
[67] *Fairfield Ledger*: June 14, 1855, p. 2, col. 6.

> Thus has ended the first case under the fugitive-slave law in Iowa. How opinions change! Four years ago, Mr. ----- and myself, and not to exceed three others in town, were the only men who dared express an opinion in opposition to the fugitive-slave law, and, because we did express such opinions, we were denounced like pickpockets. Now I am Governor of the State; three-fourths of the reading and reflecting people of the county agree with me in my sentiments on the law, and a slave could not be returned from Des Moines County into slavery.

While "Dick" was imprisoned, Grimes had mused that if he were a private citizen he thought he should be a law-breaker, and during the trial he had packed the courthouse with his anti-slavery friends. The penalties of the Fugitive Slave Law were severe—in sum, anyone harboring, feeding or assisting a runaway slave, or refusing to assist in a slave's recapture, was subject to a fine of up to $1,000 and imprisonment of up to six months per fugitive—but in denying any accused Black the right to a trial or any evidence other than the assertion of the slave-catcher, in paying slave-catchers for every Black they "returned" to the South, in paying commissioners $10 if they found for the slave-owner and only $5 if they found for the Black, and in requiring neutral citizens in Free States to aid slave-catchers, the very harshness of the law actually encouraged more fervent abolitionism and UGRR activity than ever before.[68]

W. W. Junkin reported the case succinctly:

> The Fugitive Slave "Dick," who was arrested at Burlington in this State last week, was discharged by the U. S. Commissioner, there not being sufficient evidence produced by the claimant to satisfy the Commissioner that the "Dick" arrested was *the* one they were after.[69]

Though Junkin chose to downplay it, Governor Grimes correctly perceived that the victory, though small in itself, revealed a deep shift in Iowa's attitudes toward slavery over the span of a mere four years. Meanwhile, the mainstream Democrats adhered to their pro-slavery agenda, apparently undeterred by the growing anti-slavery sentiment in town, in the state, and in their own ranks.

---

[68] William Salter, *The Life of James W. Grimes, Governor of Iowa, 1854-1858; A Senator of the United States, 1859*-1869, New York: D. Appleton & Co., 1876, p. 73.

[69] *Fairfield Ledger*, July 5, 1855, p. 2, col. 3.

June 30, 1855: Pro-Slavery Democratic Convention

Fairfield's mainstream Democrats ignored the Independent Democratic Convention, and about a month afterward—three months after the Kansas election fraud—the "land-office clique" led by Hon. Bernhart Henn and Col. James Thompson, past and present Fairfield land-office registers—met in a pro-slavery Democratic Convention on June 30, 1855. As a prerequisite for admission to the convention, every delegate had to stand up before lawyer Samuel Clinton and "pledge that he was a Democrat, and had no sympathy with the Know Nothings." A secret order arising in reaction to the sudden influx of large numbers of Irish and German immigrants into the United States in the 1840s and 1850s, the Know Nothings were anti-foreigner and anti-Catholic, and often also pro-temperance—opposing the Irish predilection for whiskey and the German predilection for beer—and anti-slavery. Many who left the disintegrating Whig Party became Know Nothings, at least briefly, and when the Know Nothing Party later adopted a pro-slavery platform, many disaffected Know Nothings became Republicans.

In the Democratic Convention, *Sentinel* editor David Sheward moved to appoint a committee to present resolutions, and was appointed chairman of that committee. One of his resolutions "required that every candidate for nomination" again, "come forward and pledge himself that he was not a Know Nothing and would not join that order during his term of office" if nominated and elected. A small majority amended the resolution, over the objections of the clique, to allow an absent man's friends to attest in his behalf. W. W. Junkin called the whole resolution a farce, as being sworn to secrecy, a Know Nothing could neither admit nor credibly deny that he was one; nor could his friends attest he wasn't. Interestingly, the resolution singled out the Know Nothings but did not mention the Sag Nichts, a secret society of immigrants organized to oppose the Know Nothings; the Administration favored the Sag Nichts. Junkin lampooned the "thumb-screw regulations" of the "Holy Inquisition" and said the clique formulated the original resolution to weed out contenders like the absent McGaw and leave only the nominees they had decided to back long beforehand: Thomas McCullough for judge, and Thomas Shamp for recorder and treasurer. Both these men were duly nominated and indeed elected, as was the Democrat Samuel Clinton for prosecuting attorney, succeeding his law-partner, the Republican Caleb Baldwin.[70]

One of Sheward's resolutions took aim at both the Nativist Know Nothings and those opposing slavery by branding them un-American:

> That they are not all true Americans who are born in America: for among them are Monarchists, Federalists, Fanatics, secret plotters, unprincipled demagogues, and all those who would sacrifice their country's prosperity and freedom for their own temporary success.

Another resolution lauded the Democratic Party as that of the "true American" in upholding the Constitution, extending freedom to the oppressed, and despising all tyranny:

> *That he only is a true American who loves the principles of Democracy*, adheres faithfully to the Constitution of the United States, labors to extend the principles of free government throughout the world, and to the oppressed everywhere, and cordially despises 'every species of tyranny over the mind of man.'

Passing over the Democrats' recent "free-government principles" of pro-slavery election fraud in Kansas and their oppression and tyranny over Blacks at home, Junkin exposed the Democratic clique's "secret plotting" and tyrannical treatment of its own party-members, as well as its attempts to stir up hatred and divisiveness:

> Every man is not a true Democrat who makes loud professions of Democratic holiness. True Democracy consists not of a ruling of the many by the few—as in the case where a few leaders, by plotting behind the scenes, forces upon the masses of a party intolerant resolutions, filled to repletion with everything but the truth. It is not the office of true Democracy to teach the citizens of this great Republic to hate each other, and to stir up discord among the common brotherhood of the country by branding as aliens ... all whose minds do not harmonize with the diseased imagination of the man who manufactures party resolutions to order.

And W. W. Junkin concluded with a passionate plea:

---

[70] *Fairfield Ledger*, July 5, 1855, p. 2, cols. 1-2.

> Let us ask the honest Democrats of Jefferson a question, Do you believe that your neighbor who differs with you in political sentiment is a "Monarchist," a "Federalist," a "Fanatic," a "secret plotter," an "unprincipled demagogue," or "one who would sacrifice his country's prosperity and freedom for his own temporary success," merely because he does so differ with you? Do you believe that? We will not believe you do, unless you ... endorse the hateful, intolerant resolutions which a few leaders have forced upon you and now demand you to support.
>
> A few politicians by clique associations may crowd such resolutions through a convention; but we doubt whether the liberality now abroad amongst the people will sustain such clique management, and such intolerant resolutions. The people have been rode to death by clique management. Will they stand it longer? We shall see.[71]

Junkin would soon reveal that the preamble and resolutions of the Democratic Convention of June 30 had already been drafted in a primary on June 28 by D. Sheward, Hon. B. Henn, Col. James Thompson, Dr. J. C. Ware, and S. H. Bradley—with an aside: "[Do the names look anything like the Land-Office clique?]"—who had received them from old stalwart Democratic Judge Joseph Knapp of Keosauqua, who had given them at a Van Buren County Democratic Convention which the *Sentinel* had quoted on June 21. Junkin cited the Democrat Delazon Smith, who had quoted Judge Knapp as once saying:

> Why... we go for our own interests! No anti-Dodge man shall have office! We have built up the Democratic party, and given it all the character and respectability it ever had in the country! No other man can or shall go to the Senate, but A. C. Dodge! This Jefferson Democracy is a *great humbug! Talk of the intelligence of the people! Look at them! One half of them can neither read nor write! They are just like a flock of sheep—wool will grow upon their backs, and if by calling them clever fellows, and intelligent yeomanry, and sovereigns, we can get them to stand still long enough for us to take the fleece off, we are so much smarter than the Whigs!—Give me three or four men in each county seat, and I have no reason to trouble myself about the wool shirt ignoramuses! All you have got to do when all is arranged is to whistle, and they will come! Just shout thunder, blood, Democracy and General Jackson, and they will go it with a rush!*"

---

[71] *Fairfield Ledger*: July 12, 1855, p. 2, cols. 1-2.

Junkin pointed out that the "Sham Democracy" was still following Knapp's elitist formula, and that Wapello, Monroe, and other counties to the west had also copied Knapp's resolutions verbatim, doubtless through his contact with "three or four men in each county seat."[72]

While Junkin's revelation about Knapp appeared in the *Ledger* after the People's Republican Convention on July 14, the Republicans had seen enough of the Democratic machinations to condemn them in their own convention.

July 1855: People's Republican Convention in Fairfield

A call for a People's Republican Convention to be held in Fairfield on July 14, 1855 garnered 111 signers in the *Ledger*, with others whose names were illegible:

> Peoples' Republican Convention. The undersigned citizens of Jefferson County impressed with the importance of a faithful, honest and capable administration of our county offices, and feeling it a duty we owe ourselves and the public, to oppose by every fair and honorable means the ticket recently proposed by the so-called "Democratic Convention," respectfully suggest to the voters of Jefferson county to hold a People's Republican Convention in Fairfield on the 14th Day of July, at 10 o'clock A.M., and we propose as our platform – an upright administration of our county offices – and the only requirements that should be exacted of the candidates that they should be honest and capable.
>
> E. S. Gage, J. A. McKemey, Matthew Clark,
> J. Robertson, E. Billingsley, Jas McCullough
> J. L. Myers, W. P. Winner, J. W. Messick,
> T. D. Evans, J. V. Myers, J. M. Runnells,
> Joseph Ball, E. C. Hampson, Jos. Junkin, Sr.,
> A. Fulton, Eli Hoopes, J. M. Whitham,
> A. T. Wells, W. H. Darling, R. F. Ratcliff,
> Jas. Wamsley, Wm. Bolding, Peter Slimmer,
> W. T. Campbell C. Baldwin, W. Dunwoody,
> W. S. Parker, B. F. Freeman, J. E. Mobley,
> A. P. Littleton, G. W. Pancoast, S. Kirkpatrick,

---

[72] *Fairfield Ledger*: July 19, 1855, p. 2, col. 2; July 26, 1855, p, 2, cols. 1-2.

W. Vaught, W. W. Junkin, I. Armstrong,
N. R. Immell, W. P. Irland, G. M. Chilcott,
L. H. Parsons, Joseph Dole, Geo. Cochran,
Jeff'son Cook, Wm. Harper, J. O. Kirkpatrick
S. D. Parker, W. S. Cooke, S. W. Taylor,
J. W. Cochran, H. Gorsuch, Stephen Butler,
W. T. Day, Wm. S. Layton, Jn. McCullough
W. S. Lynch, John Cochran, Moses Black,
H. P. Warren, C. W. Slagle, Colman Graves,
C. E. Noble, C. T. Robinson, Geo. Craine,
J. H. Wells, E. Mechem, G. A. Smith,
T. B. Fleanor, N. H. Campbell, Wm. Bickford,
D. Young, John Fleenor, J. P. Fansher,
Hiram Gregg, Henry Frush, W. B. Campbell,
B. B. Tuttle, Wm. Miller, J. C. Fetter,
Abaha Salins, G. W. Robinson, J. W. La Force,
John Roberts, Robt. Pleugh, T. M. Thompson,
J. A. Ireland, H. A. Robinson, R. R. Hall,
A. Hughes, E. Darling, S. M. Northup,
Steph Martin, Sam'l Robb, L. B. Hughes,
John Rider, James Jeffers, Jacob Snook,
S. B. Burgese, Swain Hanes, H. D. Kness,
W. H. Walker, J. T. Moberly, Joseph Blakely,
M. Wilson, Jacob Gower, J. H. Walden,
G. D. Milligan, Grinder Wilson, A. R. Fulton,
S. P. Majors, George Fisher, Goodman Graves.

Above we give place to a call for a People's Republican Convention. We have been compelled to omit some of the names which were signed to the call, owing to our inability to decipher the signatures. We hope that the people of old Jefferson will remember the 14th inst., and come up to the Convention in their strength. Let us have an expression of the voice of the people.[73]

The convention of 176 voters met at the court-house and carried Dr. J. L. Myers' motion to call C. W. Slagle to the chair, Dr. Mechem's motion to choose A. R. Fulton as secretary, and S. P. Majors' motion to

---

[73] *Fairfield Ledger*, July 12, 1855, p. 2, col. 1.

appoint a committee of one man from each township to select permanent officers and draft and report rules. Chairman Slagle appointed S. P. Majors of Liberty, J. L. Myers of Fairfield, A. W. Langdon of Cedar, S. Robb of Locust Grove, John Williams of Penn, W. S. Cook of Round Prairie, E. Mechem of Polk, John Park of Walnut, J. H. Hendricks of Lockridge, Robert Wilson of Des Moines, and Richard Gaines of Black Hawk. (Buchanan Township's administration was divided equally between Fairfield and Lockridge Townships until 1856, and so was missing from Junkin's report.)

On Caleb Baldwin's motion, C. W. Slagle appointed Caleb Baldwin, Matthew Clark, J. S. Mount, John W. DuBois, and G. Hanawalt as a committee to draft resolutions expressing the convention's sense.

After reconvening in the afternoon, the township committee reported the convention's rules of attendance, nominations, voting, speaking, and rebutting, and named C. W. Slagle as the convention's president, with Thomas Moorman and L. B. Hughes as vice presidents, Ebenezer S. Gage as secretary, A. R. Fulton and J. W. DuBois as clerks, and B.B. Tuttle and C. C. Collins as tellers.

Caleb Baldwin as chair of the five-man committee reported eight resolutions repudiating party control and opposing slavery's expansion, declaring:

1. They had implicit faith in a Republican form of government;

2. A voter was responsible to his country, not to any political party;

3. A voter had an "absolute right" to form his own political opinion, and it was anti-republican for any party or persons to control the voter;

4. They refused to endorse any political parties, but recognized the "common brotherhood laboring for the common good of the country," repudiating the right of any party to brand others of "the great brotherhood" as "Monarchists, Fanatics, Traitors, or villains;"

5. The people's voice was supreme in selecting public officers;

6. They unreservedly endorsed "the Republican qualifications for office—Honesty and Capability;"

7. "On the subject of slavery, this proposition: Shall freedom be confined to the free States, or slavery to the slave States? As the sense of this convention, we pronounce the latter;"

8. They would "use every honorable means, as a free and independent people," to secure their candidates' election. That day they nominated David J. Evans for county judge, Henry P. Warren for recorder and treasurer, G. M. Chilcott for sheriff, and G. W. Pancoast for coroner.

C. E. Noble then called up four more resolutions previously introduced by Richard Gaines:

1. The "aggressions of slavery," especially the "Nebraska outrage" and the Kansas election fraud, had "aroused the freemen of the Republic," and they would "maintain their rights and resist the addition of slave territory."

2. They would "maintain the nationality of freedom."

3. The "friends of freedom should make principles, not birth-place, the test of admission to citizenship."

4. They would "repel every ecclesiastical interference in political affairs, by Potentate, Pontiff, or Priest, as destructive of the right to worship God according to the dictates of conscience and liberty."

After some discussion, the convention tabled Gaines' resolutions on L. B. Hughes's motion—it was wrongly rumored that the Free-Soil Democrat Gaines had colluded with the Democratic land-office clique to introduce these "firebrands" to sow dissension—and adjourned. But Fairfield's new Republicans had found common ground: resistance to control by either party, Whig or Democrat, and opposition to slavery's expansion.[74]

Thus was unofficially born Jefferson County's Republican Party in Fairfield, on July 14, 1855. Though beginning as a third-party coalition against the expansion of slavery, and weak at first—none of their Jefferson County nominees of 1855 was then elected—the Republicans would go on to nominate John C. Frémont for President in 1856, and to win with President Abraham Lincoln in 1860.

Iowa's Republicans originally considered holding their first state convention during the second Iowa State Fair in October 1855 in Fairfield. The second Iowa State Fair continued the first fair's combination of sheep and abolitionism. Taking first prize for his long-wool ewes was William Clawson of Lee County, Iowa, whose father-in-law Joel Parker was an abolitionist editor running a store in Indiana which offered only free-labor goods. Joel Parker's business partner was Indiana's UGRR "President" Levi Coffin.

Though primarily showing Iowan goods—including cattle, oxen, horses, mules, sheep, poultry, farm implements, produce, cloth, prepared foods, artwork and inventions—the fair also attracted first-rate contestants from across the country. Taking diplomas for both first and second prize for best cultivator was one John Deere from Illinois, and

---

[74] *Fairfield Ledger*, July 19, 1855, p. 2, cols. 6-7.

the Committee recommended a diploma to "Sharp's American Breech Loading Rifle, manufactured at Hartford Connecticut by Sharp's Rifle Manufacturing Co.," calling it a "neat and curious gun."[75]

The quick-loading, highly-accurate, long-range Sharps rifle was the weapon of choice for John Brown and New England emigrants to Bleeding Kansas of 1855-56, and was still favored by U. S. Army sharpshooters and by the cavalry of both sides in the Civil War.[76]

In the end, Iowa's Republicans decided not to mix agriculture and politics, and eventually followed a recommendation by Richard Gaines on December 9, 1855 in the *Ledger* to hold their first convention on February 22, 1856 in Iowa City.

1855-1856: Fairfield Awakens to Bleeding Kansas

Much of Jefferson County's slowly-growing opposition to slavery came from *Fairfield Ledger* editor W. W. Junkin's incessant struggle to educate his readers. While he battled with the *Sentinel* and ran anti-slavery pieces in every issue of the *Ledger*, between November 15, 1855 and January 17, 1856, he methodically laid out a numbered series of seven articles on "Why we are opposed to the Extension of Slavery," which must have had a cumulative and powerful effect on any undecided.

On February 22, 1856, the preliminary National Republican Convention met in Pittsburgh, Pennsylvania, chaired by Kentucky's Free-Soil Democrat and Republican statesman, Francis P. Blair. On that same day, Wickliffe M. Clark and A. R. Fulton were delegates from Jefferson County to the Republican State Convention in Iowa City.

After the Iowa City Convention, Jefferson County Republicans met at Fairfield's court-house on March 15, 1856 to ratify the Iowa City

---

[75] *Report and Proceedings of the Iowa State Agricultural Society, Held at Fairfield, Iowa, in October, 1855*, Fairfield, Iowa: Iowa Farmer and Iowa Sentinel Offices, 1856, pp. 18, 30.

[76] Some suppose that the term "sharpshooter" derives from "Sharps' shooter," but Michael Quinion notes that Christian Sharps first offered his rifle for sale in 1850, nearly a half-century after the *Edinburgh Advertiser* of June 23, 1801 described the British Army's new "Sharp Shooters:" evidently, directly from the German *Scharfschütze, recorded in* Jacobsson's *Technologisches Wörterbuch* of 1781. See Michael Quinion, *World Wide Words*, "Sharpshooter," page created April 14, 2007: http://www.worldwidewords.org/qa/qa-sha5.htm

platform, organize the county, and nominate a Republican county ticket. They appointed Benjamin Robinson as president and UGRR operator C. O. Stanton as secretary, and one person from each township as a committee to prepare business, while James F. Wilson addressed the meeting. After a "short absence," the committee reported:

> United in a common resolve to maintain Right against Wrong, and believing in the determination of a virtuous and intelligent people, to sustain Justice, we declare,
>
> 1. That governments are instituted among men to secure the inalienable rights of life, liberty, and the pursuit of happiness.
>
> 2. That the mission of the Republican party is to maintain the liberties of the people, the sovereignty of the States, and the perpetuity of the Union.
>
> 3. That under the Constitution, and by right, Freedom alone is National.
>
> 4. That the federal government being one of limited powers derived wholly from the Constitution, its agents should construe those powers strictly, and never exercise a doubtful authority—always inexpedient and dangerous.
>
> 5. That if this plain Jeffersonian and early policy was carried out, the federal government would relieve itself of all responsibility for the existence of slavery, which Republicanism insists it should, and means it shall do, and that regarding slavery in the States as a local institution, beyond our reach, and above our authority, but recognizing it as of vital concern to every citizen in its relation to the nation, we will oppose its spread and demand that all territory *shall be free.*
>
> 6. That the repeal of the Missouri Compromise, and the refusal of the slave power to abide by the principle on which that repeal was professedly based, make the national domain the battle ground between freedom and slavery, and while Republicans stand on a national basis, and will ever manifest and maintain a national spirit, they will shrink from no conflict, and shirk no responsibility on this issue.
>
> 7. That the slave power, the present national Administration, and its adherents, having violated this policy, and the principle on which it is based, by a disregard of law and its own profession, by an invasion of the state and personal rights, and by breaking solemn covenants has forced upon the country the Issue, whether freedom shall be limited to the free States, or slavery to the slave States, and made that issue absorbing and paramount.

After thus ratifying and affirming "the above Republican platform as adopted at the Republican State Convention" at Iowa City, the Jefferson County Convention added that they sought no unity of opinion or belief on minor matters; that merit, not birthplace should be the test for office-seekers, applying only Jefferson's rule: "Is he honest? Is he capable?" And that they should "welcome the exiles and emigrants from the old world, to homes of enterprise and of freedom in the new." Also, as Iowan slavery extensionists were reporting that the Republican State Convention's nominees had pledged to support the Philadelphia Know Nothing Convention's presidential and vice-presidential candidates, who stood on a pro-slavery platform, they called on the nominees to confirm or deny those reports so that the Republican Party could present true Republican candidates to the people. They also resolved that the delegates give Secretary Stanton the names of three persons from each township to act as committee men, and that Republican clubs be formed in each township as suggested by the Republican Association of Washington City, with the central executive committee to furnish that association's organizational form to each township. Finally, each candidate for nomination should declare his adherence to the Republican principles.

Robert F. Ratcliff, W. M. Clark, J. L. Myers, W. W. Junkin, and E. C. Hampson were chosen as the convention's central committee. For the Republican county ticket, Jacob H. Allender emerged as nominee for school-fund commissioner with 22 votes, other candidates being Archer Green (7 votes), Joseph Junkin Sr. (4), and H. P. Warren (1). The convention chose Wickliffe M. Clark as their nominee for sheriff with 20 votes over Solomon Stever (8 votes), W. B. Clapp (3), R. H. Van Doren (2), John Rider (1), and Johnston Moore (1).

The convention recommended committee men and requested them to organize Republican Clubs in their township, naming R. Gaines, H. Hutchins, and S. P. Cushman for Black Hawk; J. W. Green, W. Collins, and W. D. Clapp for Lockridge; W. H. Copeland, John Winsell, and A. O. Edwards for Des Moines; John Spielman, Lewis Rader, and H. Gorsuch for Walnut; and Isaac Ellis, UGRR operator C. O. Stanton, and "E. E. Osborne" [T. E. Osborn] for Penn. The convention then adjourned.[77]

Meanwhile, Missouri's pro-slavery Border Ruffians were aggressively attacking, robbing, and repelling Free-Soil settlers who were attempting to reach Kansas through their state, and so more and more Free-Soilers were avoiding Missouri altogether by traveling across Io-

---

[77] *Fairfield Ledger*, March 20, 1856, p. 2, col. 4.

wa into Nebraska and then south into Kansas. Fairfield's pro-slavery *Sentinel* editor took issue with their passage, calling them "traitors"— the old epithet which pro-slavery politicians had long wielded so effectively to intimidate their anti-slavery foes. W. W. Junkin was not so easily intimidated, and he did not spare the slave-culture at home any more than in Missouri; on April 10, 1856, he published an article titled, "Border Ruffianism in Iowa:"

> Judging from the tone of an editorial in last week's Sentinel, we should conclude that the editor of that paper is growing somewhat jealous of his pro-slavery allies of the Missouri border. He seems disposed not to let them carry off all the laurels from the fields whereon bad men fight the battles of slavery, to secure its establishment in Kansas. He grew quite indignant at the idea of Kansas emigrants passing through Iowa, on their route to their new homes, where they intend to struggle for Freedom, and cried out: "Shall they be permitted to pass through our State?" and then launched forth his venom about "traitors" and "treason."
>
> We have little doubt but that the editor of the Sentinel thinks it a very treasonable thing for men to hate the cursed institution of slavery, which he seems to love so much, and so greatly desires to protect, extend and perpetuate, and he may be very willing and anxious to get up a spirit of border Ruffianism in Iowa, in order to close the highways of our State against those who may wish to traverse them while on their way to new homes in persecuted Kansas, but he will find that the people of Iowa will have little sympathy with his border ruffianism. The highest crime he can lay to the charge of the Free State emigrants to Kansas is, that they desire to establish the principle of freedom in the West on as firm a foundation as it rests on in the East. For this he asks, Shall the traitors be permitted to pass through our State? He may rest assured that they will be permitted to pass through Iowa. The people of Iowa sympathize with the Free State men in Kansas, and with all who may go to that territory to settle and cast their lots with the free State men, and they will not only permit them to pass through this State, but lend them aid if aid be necessary to secure to them their rights.
>
> Iowa is a poor market for the border ruffianism of the Sentinel. Such stuff would pass more freely and currently in the western part of Missouri, where men reason with pistols and bowie-knives.
>
> We suppose the editor of the Sentinel thought that he must do something to carry out the new doctrine of the Administration party, which Douglas proclaimed in the Senate when he declared: "Sir, we

mean to subdue you!" but he will find it an up-hill business—it won't work.[78]

After the Border Ruffians had intimidated, expelled, and killed Free-Soilers with relative impunity in Kansas for nearly two years, the citizens of the Free-State town of Lawrence resisted an arrest attempt by the pro-slavery Sheriff Jones, shooting and slightly wounding him in the process. On May 21, 1856, Sheriff Jones returned with cannon and 800 armed men, waving the state flags of Alabama and South Carolina, and flags with the mottos "Southern Rights," "Supremacy of the White Race," and "Kansas the Outpost."

The citizens submitted peacefully, but the pro-slavery army sacked Lawrence, burning their pre-eminent free-state leader Charles Robinson's house, destroying the Free State Hotel and the newspaper offices of the *Kansas Free State* and the *Herald of Freedom,* and looting the town. The next day Democratic Representative Preston Brooks of South Carolina beat Republican Senator Charles Sumner of Massachusetts bloody, unconscious, and almost to death with his cane in the U.S. Senate chambers: the punishment the Southern "code of honor" prescribed for one's inferiors, in lieu of a duel between equals. In Sumner's two-day speech on May 19-20, 1856, "The Crime Against Kansas," he had lambasted the Slave Power for the "rape" of the "virgin Territory" of Kansas, "compelling it to the hateful embrace of Slavery," and had ridiculed the authors of the Kansas-Nebraska Act: Stephen A. Douglas of Illinois and Andrew Butler of South Carolina, a distant relative of Preston Brooks. Of Butler he had said:

> The senator from South Carolina has read many books of chivalry, and believes himself a chivalrous knight with sentiments of honor and courage. Of course he has chosen a mistress to whom he has made his vows, and who, though ugly to others, is always lovely to him; though polluted in the sight of the world, is chaste in his sight—I mean the harlot, Slavery…..[79]

The incident further polarized an already divided nation. Senator Sumner became a hero to Northerners, while Congressman Brooks, who had broken his gold-headed, gutta-percha cane in his relentless

---

[78] *Fairfield Ledger*, April 10, 1856, p. 2, col. 1.
[79] Hon. Charles Sumner, "The Crime Against Kansas. The Apologies For the Crime. The True Remedy," Washington, D.C.: Buell & Blanchard, 1856, pp. 4-5.

attack, received the hearty congratulations of Southerners along with hundreds of replacement canes. W. W. Junkin was quick to point out a pattern of bullying by the pro-slavery faction:

> The Administration members of Congress have been prolific during the present session, in their displays of their characteristic border ruffianism. The first exhibition of this kind was the cowardly attack of Rust of Arkansas, on Mr. Greely; then followed the cold blooded murder of the Irish waiter at Willard's Hotel, by Herbert of California; and now we have accounts of a most dastardly and brutal attack on Senator Sumner by Brooks of South Carolina.—These ruffians are all warm supporters of the Administration, and are practising in Washington what Pierce's minions are practising in Kansas. The disgraceful exhibitions of ruffianism are all confined to the Administration supporters, who, being engaged in a bad cause, leave argument to resort to cains [sic], pistols, bowie knives and other weapons of bullies, cowards and blackguards.
>
> Terrible times, indeed, have fallen upon us, when the mere expression of opinion and an assertion of the right of free discussion are to subject Northern Senators to brutal assaults in their very seats in the Senate chamber. If the South intends to enforce her infernal doctrines upon the North by striking down and assassinating every Northern man who has the courage to show a front against the steady encroachments of the debasing institution of slavery, the sooner the people of the North understand the matter the better. If United States bayonets are to conquer the free spirits of Kansas, and force upon its oppressed people the detestable code of laws enacted by the bogus legislature of that Territory; if the bullies whom the South sees proper to send to Congress are to turn assassins, to beat and murder Northern men; if mob law and brute force are to be resorted to, to silence freedom of speech, all for the sake of extending slavery and securing Southern votes in support of the reckless demagogues who are now controlling the destinies of the Administration party, we hope the country may soon understand the matter, and then, come weal or come woe, let those who have defend themselves.
>
> If the bullies of the South are permitted to go on for a few years more as they have been going on for the past year, what better will our country be but a despotism?—Northern men will be dogs if they submit to such insolence longer. We would not counsel retaliation by brute force; but we do appeal to the North to send to Congress a united body of men opposed to the aggressions and ruffian insolence of such scape-gallows' as are now piloting Southern arrogance throughout every artery of our national body. It is not worth while depending

on supporters of the Administration: for they are all banded together to support the one grand aim of the South—the subjugation of the North. The North must unite on the right kind of men or be content to be conquered. This is plain from the many acts of aggression, ruffianism and bloodshed constantly emanating from Southern forces. If such things were of rate occurrence, they might be overlooked; but when they follow in such quick succession as they have for a few months past, they admit of no other interpretation that that of a settled determination on the part of the South to force the North into measures by resorting to any means, not even stopping at bloodshed and murder.

We have already seen the conduct of Brooks excused in Administration journals. Let them take the benefit of their attempted excuse. The man who would volunteer in such a defense is no better than the assassin whom he defends, and if such men expect the people of the North to support them, we have faith enough in the uprightness of the people to believe that such expectations will be most emphatically rebuked.[80]

But even as Junkin wrote in outrage and despair, the North had found its hero, a man willing to stand up to the "bullies of the South" in a way they could understand. On May 24-25, 1856, John Brown and a band of abolitionist settlers killed five pro-slavery settlers near Pottawatomie Creek. Southern-sympathizing historians would later revile John Brown as a madman and a murderer, but as Missouri Border Ruffian and Confederate General Joseph O. Shelby later admitted, "I was in Kansas at the head of an armed force [in 1855-56]. I was there to kill Free-State men. I did kill them." He continued, "No Missourian had any business there with arms in his hands. I ought to have been shot there, and John Brown was the only man who knew it and would have done it." He added, "I say John Brown was right. I knew the men he killed. I condemn his killing of the younger Doyles [at Pottawatomie Creek], but the others got only what they deserved. After that I had great respect for Old John Brown." He concluded that John Brown "was the only Kansas man who had the right idea of the conditions existing there and the only man who had the courage to resist Missourians at the muzzle of the rifle."[81]

---

[80] *Fairfield Ledger*, May 29, 1856, p. 2, col. 1.

[81] William Elsey Connelly, *Quantrill and the Border Wars*, Cedar Rapids, Iowa: The Torch Press, 1910, pp. 288-289, cited in Lowell J. Soike, *Busy in the Cause: Iowa, the Free-State Struggle in the West, and the Prelude to the Civil War*, Lincoln,

1855-1859: *Local Iowans and John Brown*

As mentioned, John Brown's fellow sheep-judge Benjamin C. Perkins had brought his sheep from Ohio to the Iowa State Fair at Fairfield in October 1854, perhaps in part on a scouting mission for Brown. Both B. C. Perkins and his third cousin Simon Perkins, John Brown's sheep partner, had a prominent if distant cousin in Fairfield who was an abolitionist himself, with other strong family ties to both woolen manufacturing and to the abolitionists who viewed sheep's wool as a Free-Soil alternative to slave-raised cotton. Fairfield's Congregationalist Rev. Charles Gates was son of Asa Gates, a Massachusetts woolen manufacturer. Asa's third cousin Thomas C. Perkins married Mary F. Beecher, sister of Harriet Beecher Stowe—whose *Uncle Tom's Cabin* sparked anti-slavery sentiment nationwide—and of Henry Ward Beecher, who supported arming Free-Soil settlers with Sharps rifles. Thomas and Mary Perkins' daughter Emily—Asa's fourth cousin, B. C. Perkins's second cousin—married Rev. Edward Everett Hale, abolitionist supporter of a Free-Soil Kansas and co-founder of the Massachusetts Emigrant Aid Company. Asa's fourth cousins included Simon Perkins, wool-partner of John Brown, and Benjamin C. Perkins, whose sheep won the awards at the first Iowa State Fair in Fairfield in 1854.

Between 1855 and 1859, John Brown crossed and re-crossed Iowa en route to and from Bleeding Kansas, where his sons had settled in the spring of 1855. As he did, he both strengthened UGRR lines across Iowa, and inspired local Iowans to help make Kansas a free state. Henry Ward Beecher, Charles Gates' cousin's in-law who publicly advocated supplying Free-Soil settlers with Sharps rifles, was, like Gates, an anti-slavery Congregational minister, and like Beecher, Rev. Gates would soon also be accused of providing Sharps rifles to Free-Soil Kansas immigrants. The UGRR Byrkit family's pastor, Rev. Levin B. Dennis, left Fairfield in 1855 to become the first Methodist minister in the Free-Soil town of Lawrence Kansas.

About 14 miles north of Fairfield in the UGRR station of Richland, several Quaker families responded to Bleeding Kansas. In 1856 Nelson King and his brother-in-law Jesse B. Way—son of Salem's UGRR op-

---

Nebraska: University of Nebraska Press, 2014, p. 174 and p. 253, n. 114. A brilliant Confederate general in the Civil War, Joseph O. Shelby refused to surrender in 1865 and led his troops to Mexico, where they colonized for two years until Emperor Maximilian's death. Shelby then returned to Missouri to farm. He was U.S. marshal of Western Missouri from 1893 until he died in 1897.

erator Paul Way—both took their families to Osawatomie Township, Miami County, Kansas, where as "renegade" Free State men they associated with "Old Osawatomie" himself, John Brown. Jesse became sheriff of neighboring Franklin County in 1857, but was soon stabbed to death, reportedly in a fight over slavery.[82]

About seven miles east of Richland in Clay, UGRR operator Manning B. Mills was John Brown's second cousin. In 1857, Manning's brother Oliver Mills moved from Denmark, Iowa to become a UGRR station master near Lewis, Iowa, where he and ex-Fairfielder and fellow UGRR station master George B. Hitchcock reportedly hosted John Brown.[83]

About 20 miles east of Richland, 17 miles north of Washington, the notable UGRR station of Crawfordsville had already hosted the Free-Soil convention in 1854 which helped elect James W. Grimes governor. In 1856, a committee sent out armed Free-State settlers from Crawfordsville to Kansas. As late as 1899, some remembered:

> The infamous Kansas-Nebraska bill had passed and Stephen A. Douglas had set forth his doctrine of squatter sovereignty which allowed new states to decide whether they would be free or slave.
>
> The riffraff from Missouri and Arkansas went across the border and voted in Kansas, so that the proslavery vote was in the majority. The states of Massachusetts, New York and Connecticut established immigrant associations and gave bounties to settlers ... so that when the election came off it would decide against slavery.
>
> The citizens of Crawfordsville were strongly anti-slavery and they offered inducements to settlers that would go to Kansas. R. T. McCall gave five dollars, but the writer does not now know who else contributed to the fund.

---

[82] Find A Grave: Jesse Britton Way, https://tinyurl.com/jessebway. For Jesse B. Way and Nelson King as "renegade" free-state men, see H. H. Williams's letter of Oct. 12, 1857 to John Brown, cited in James C. Malin, op. cit., pp. 706-707, and n. 71. William Hutchinson of Lawrence had issued J. B. Way and Nelson King one Navy Revolver each: F. B. Sanborn, *The Life and Letters of John Brown: Liberator of Kansas, and Martyr of Virginia*, Boston: Roberts Bros., 1891, p. 366.

[83] Manning and Oliver were sons of Harlow Mills and Faith Ann (Spencer), daughter of Roswell Spencer and Faith Ann (Mills), daughter of Elizabeth (Higley) and Gideon Mills, whose son Gideon Mills married Ruth (Humphrey) and had Ruth Mills, who married Owen Brown and had John Brown. For Oliver Mills as host to John Brown, see "Oliver Mills," *Annals of Iowa* vol. 8, no. 3 (1907), pp. 237-238, online at: http://ir.uiowa.edu/annals-of-iowa/vol8/iss3/18.

Five young men offered themselves in 1855. They were James and Thompson Crawford, Lee Glenn, James Allen and Tom Smith. They armed themselves with Colt's revolvers and bowie knives and looked very savage to the small boys around. They fitted up a prairie schooner and camped on the branch, south of town, where they practiced at a mark.

In time they got ready, the people bade them good-bye and they wound their way outside the border of Missouri, as the citizens would not let them pass through that state. These men were all brave and were under the command of old Osawatomie Brown.[84]

James and Thompson Crawford's sisters, Sarah and Melissa, also served in Bleeding Kansas, and were "both proficient" with the Sharps rifle, as well as in making cartridges. They stocked and watched over their brothers' Kansas claims while the men fought "the border ruffian war," and passed "many sleepless nights" on the "wild prairies," where "with rifles in hand," they kept a "lonely vigil ... with no help within miles should they be attacked, and as they then hourly expected."[85]

About 40 miles north-northeast of Crawfordsville in Springdale, Iowa, John Brown trained his followers in 1858 for raids on Kansas slaveholders and later for the raid on the government arsenal at Harper's Ferry, (West) Virginia, an ill-fated attempt to spark a slave uprising. Among his men two Quaker brothers named Edwin and Barclay Coppoc. Edwin was hanged with John Brown in (West) Virginia; Barclay escaped and went on to fight in the Civil War, but was killed in Missouri in 1861 when his train was derailed on the Platte River Bridge

---

[84] *Washington Gazette*, May 12, 1899, p. 6, col. 2. The article continues, "What became of Tom Smith no one knows. Lee Glenn went into the army and now lives in Kansas City, but never returned to this county. James Allen returned and served in the 25th Iowa. He is now lying at the point of death in Illinois." James Crawford returned and served in Co. H, 2nd Iowa. "After the war he graduated at Rush Medical College, went to California, married and long since died. Thomps Crawford started south with John Brown, but took sick and was left at Iowa City or he would have been at Harper's Ferry." He returned to Crawfordsville in 1861, enlisted in Co. C, 8th Iowa, was captured and served 29 months in Southern prisons, "perhaps longer time than was served by anyone else." Sometime after the war he "went to Arkansas and was not heard of for years, until his family was notified of his death."

[85] *Portrait and Biographical Album of Washington County, Iowa...*, Chicago: Acme Pub. Co., 1887, p. 295. The article says a committee sent out the Crawford brothers in 1856; they "became interested in the border ruffian war in Kansas, and were intimately associated with John Brown, which continued until the breaking out of the war of the Rebellion...."

by Confederate sabotage. Edwin and Barclay were second cousins once-removed of Fairfield's renowned midwife and UGRR station-master, Rachel (Coppock) Pierce

1856: Fairfield and Bleeding Kansas, the Undeclared Civil War

By 1856, most Southerners ignored the subtler anti-slavery distinctions between Free-Soiler, abolitionist, and UGRR operator, and indiscriminately called all Free-Soilers "dirty Abolitionists" and "Negro Stealers." And indeed by the later 1850s, the pro-slavery faction's aggressive violence pushed many neutral and even some pro-slavery settlers and onlookers to become Free-Soilers, and many Free-Soilers to become admitted abolitionists, and probably more abolitionists to join the Underground Railroad.

> By 1855 the general sentiment in Jefferson County had changed from pro-slavery to a sympathetic willingness to help the fugitives. After the decision in the Forrester case..., free Negroes looked on Jefferson County as a haven and a number of them came to Fairfield....
>
> The fugitives found many friends in Fairfield, for nearly every house had a barn with a haystack close by. It was not an uncommon thing to see a black face peer anxiously from a barn door, or peek out from a haystack to see if the coast were clear. The homes of the free Negroes who lived in Fairfield were also stopping places. White people could usually tell if their neighbors had out-of-town visitors. If the house was dark and there was no response to knocking, it meant they were not at home to anyone that night.
>
> People in this vicinity who remembered the Underground Railroad activities said it was impossible to estimate the number of fugitives who passed through the town at any given time, but they became so numerous that the white people were at last forced to refuse to aid them. There seemed to be no season for migration either, for the fugitives cared little for weather or temperature if a chance for freedom came.[86]

Informing Fairfielders of Bleeding Kansas firsthand was their own Reverend Levin B. Dennis, who had left Fairfield in 1855 to become the first Methodist minister of Lawrence, Kansas. On January 22,

---

[86] WPA/FWP, "The Underground Railroad," *The Negro in Iowa*, p. 82, unpublished MS, State Historical Society of Iowa, Des Moines, copy in Iowa City.

1856, Rev. Dennis wrote from Kansas to Cincinnati's *Western Christian Advocate*, pleading for aid from the East. He described the Missouri Border Ruffians' interference with elections, destruction of Leavenworth's Free-State *Territorial Register* press, and brutal hatchet-murder of Reese P. Brown, a wool-dealer from Ohio and a fiery Free-State party leader. Rev. Dennis' letter was picked up by Keokuk's *Gate City* and then reprinted on page one of the *Fairfield Ledger* by the Methodist W. W. Junkin, who headlined it, "The Truth about Kansas." He described Rev. Dennis as "a gentleman formerly stationed here in charge of the Methodist Church, who was universally beloved by the people in and out of his society, and whose word no man who knows him will doubt in the least."[87]

However, even as late as May 25, 1856—four days after the pro-slavery sack of Lawrence, Kansas, three days after the brutal caning of Senator Charles Sumner, and on the very day of John Brown's massacre of five pro-slavery settlers at Pottawatomie—prominent conservative Congregationalists in Fairfield forced the resignation of their vocally anti-slavery minister, Rev. Charles H. Gates.[88]

Jefferson County's Republicans met at Fairfield's court-house on May 31, 1856 to select eight delegates to represent them at the Congressional Convention at Ottumwa on June 5. With C. E. Noble as chairman and A. R. Fulton as secretary, on J. F. Wilson's motion C. E. Noble appointed J. F. Wilson, J. M. Liggett, and John McCleary as the three-person committee which chose Eli Hoopes, John Rider, N. R. Immell, UGRR operator C. O. Stanton, W. M. Clark, W. G. Black, J. L. Myers, and Thomas Moorman as delegates.[89]

On June 28, 1856. Jefferson County Republicans met again at Fairfield's court-house and temporarily named William Patrick chairman and George A. Hobson secretary. For a credentials committee they appointed John Cochran from Round Prairie, J. Beatty from Fairfield, Silas Deeds from Walnut, H. W. Hutchins from Black Hawk, Isaac Ellis from Penn, S. D. Gorsuch from Cedar, S. Robb from Locust Grove, B. Robinson from Polk, and S. B. Clapp from Lockridge. Chairman Patrick also appointed E. C. Hampson, W. S. Lynch, S. Robb, J. Spielman, and S. S. Tipton to a committee on resolutions and permanent organiza-

---

[87] *Fairfield Ledger*, March 6, 1856, p. 1, col. 3.
[88] Diary entry for May 25, 1856, Joshua Monroe Shaffer papers, 1850-1905, Des Moines Historical Library Manuscripts (Sh 13), State Historical Society Research Center, Des Moines, Iowa.
[89] *Fairfield Ledger*, June 5, 1856, p. 2, col. 4.

tion. This committee proposed H. B. Mitchell for president and D. J. Evans for secretary; they declined, and Benjamin Robinson presided with George A. Hobson and A. R. Fulton as secretaries.

The committee endorsed the resolutions adopted by the National Republican Convention at Pittsburgh on June 18, 1856 and re-endorsed the State Convention's platform, favored a convention to revise Iowa's constitution, approved the nomination of Samuel R. Curtis for state congress and of John C. Fremont for president and William L. Dayton for vice president by the National Convention, resolved to support this convention's nominees, and favored "Free Territory, Free Speech, a Free Press, and Fremont." The chair appointed Eli Hoopes and D. J. Evans as tellers. Caleb Baldwin won the nomination for state senator by 36 votes to Joseph Ball's 7; Louis Roeder (41 votes), W. H. Copeland (31 votes), and C. O. Stanton (25 votes) won the nominations for state representatives over Joseph Ball (12), W. D. Clapp (7), A. R. Fulton (6), D. J. Evans (4), J. Cowen (4), Moses Black (3), Joseph Fell (2), C. E. Noble (1), W. S. Lynch (1), and ___ Carver (1). By acclamation, they nominated C. W. Slagle for prosecuting attorney, R. F. Ratcliff for district clerk—he won, and served through 1860—and C. E. Noble and William Bickford for representatives to replace J. Wamsley and E. Mechem, who had moved out of Jefferson County. Messrs. Noble and Bickford served in the House through 1856. On learning he had been nominated for senator, Caleb Baldwin declined and was replaced by W. M. Read of Libertyville, who won and served in the Senate from 1856 to 1859. C. W. Slagle declined the nomination for county prosecutor and was replaced by Caleb Baldwin.[90]

That July 10, Junkin republished another letter written to Chicago's *Northwestern Christian Advocate* on June 11 from Lawrence, Kansas by Levin's son, Rev. Baxter C. Dennis, then a junior preacher to the Leavenworth mission. Baxter described the sacking of Lawrence and decried the further Border Ruffian outrages. He exclaimed, "Civil war is inevitable," and begged for Northern aid in the fight against slavery.[91]

At the same time, the Republicans Jacob L. Myers, Grinder Wilson, R. R. Hall, J. V. Myers, R. H. Van Doren, and "C. C. Collings" signed the list of those still supporting the "Know Nothing" or National American Party. They decried their Order's absorption into a "sectionalist"

---

[90] *Fairfield Ledger*: July 3, 1856, p. 2, col. 7; July 10, 1856, p. 2, col. 4.
[91] *Fairfield Ledger*, March 6, 1856, p. 1, col. 3; , July 10, 1856, p. 1, col. 2.

party which nominated foreign-born candidates, and called for "the preservation of our Glorious Union from the fanatical attacks of hot heads either North or South," and for a convention on July 12, 1856. Both C. C. Collins and Jacob L. Myers ran for state representative on the American Party ticket. To these "Americans" the ardent abolitionist and Republican Samuel Robb addressed a letter to the *Ledger* saying he had been the first in the county to admit he was an "American"—i.e., a Know Nothing—but that party had already been swallowed by the sectional slave interest as evinced by Fillmore's speech. The Virginian framers of the Constitution had never mentioned slavery and had provided that the Northwest Territory be free, so that slavery, not freedom, was sectional.[92]

On July 24, 1856, W. W. Junkin reprinted a piece from David Sheward's *Sentinel* on some anti-slavery Kansas immigrants who had come through Fairfield:

> A company of about fifteen of as woebegone and Godforsaken a looking set as one would wish to see, passed through our place on Saturday, *en route* for Kansas—armed, not with the peaceful weapons of the true settler, but with murderous weapons, with which they intend, should their courage last them, to force their way into that Territory, notwithstanding the President's Proclamation forbidding them so doing.

To this Junkin responded:

> What a pity, Sheward, that you are so easily frightened.... You saw the "murderous weapons" in the hands of these few innocent Kansas emigrants. Forthwith comes the spasm—Sharp's rifles—blood—murder—thunder, mud, torn-breeches, pitch-forks, hail and fire rush upon your brain, and out pours the natural result—a stream of low, filthy, vile slander and abuse, upon the head of an innocent and unoffending clergyman, charging him with being the agent for the distribution of Sharp's rifles—basely attempting to ward off the charge, by saying you were so *informed!* Now, Sheward, you were *never* so informed. Why doubly damn yourself by deliberately telling *two lies*, when one good lie was sufficient....

---

[92] *Fairfield Ledger*: July 10, 1856, p. 2, col. 5; July 17, 1856, p. 2, col. 4; July 31, 1856, p. 2, col. 4.

Apparently, Sheward had accused Fairfield's just-resigned Congregational Rev. Charles H. Gates of distributing Sharps rifles to Kansas immigrants, as his cousin's in-law, Congregational Rev. Henry Ward Beecher, was doing from New York, hence the Sharps' nickname of "Beecher's bibles." Junkin devoted his next article to defending Gates, "a worthy, exemplary, quiet, Christian citizen, who confines himself strictly, to the walks of private life" from one of the *Sentinel*'s "meanest, dirtiest, lowest, most contemptible articles which has ever disgraced its columns." But first, he skewered Sheward's double standard:

> But, Sheward, you say it is a great aggravation that these few Kansas emigrants should, just at this time, be found on their way to that Territory, and "notwithstanding the President's Proclamation forbidding them to do so."
>
> Now, Sheward, ... please inform us about the armed South Carolinians, Georgians, and other [Southerners], who have been, and still are, going into that same Territory, "notwithstanding the President's Proclamation?" How did these armed Southerners get there? How did Major Buford's men get there" How came they to form part of Bogus U. S. Marshal Donaldson's posse at the sacking of Lawrence? Who employed them to shoot down the peaceable settlers, drive them away from their homes, burn their houses, destroy their property, and sack their towns? Can you tell?
>
> Besides, has the President's Proclamation two faces? .... One *against* the North, and one in *favor* of the South? One forbidding Northern men going to Kansas, under the pains and penalties of expulsion from the Territory, or death if they choose to remain; and one allowing southern men to go there and be mustered into a *Posse* by a renegade U. S. Marshal, for the purpose of sacking towns, and destroying property? Is this justice? Is it right?

Junkin concluded with a contemptuous challenge to Sheward:

> Sheward, as soon as the spasm gets fully off—your physical powers again restored to an equilibrium, and your mental faculties be cleared from improper bias, (if that be possible,) will you be kind enough to answer a few of the above interrogatories? We will not impose them all upon you at once. That would be tasking your powers of mind too severely. Do the best you can, therefore, and we shall be satisfied.[93]

---

[93] *Fairfield Ledger*, July 24, 1856, p. 2, col. 2.

Sheward may have ignored Junkin's questions, but more Fairfielders were listening now. The firms of W. R. Wells & Co. and Wells & Stever, who despite their anti-slavery stance had supported both newspapers equally, now "withdrew their pecuniary aid" from the *Sentinel*. They said its editor misused free speech, "violated his office and outraged public sentiment" in abusing C. H. Gates, "a gentleman whose life and deportment among us for four years needs no defense and admits of no defamation." They added, "If there was anything wrong in the eyes of the law, either human or Divine, in giving those emigrants a dinner, others were equally culpable, and why not denounce them all." But no, they claimed Sheward abused Gates thinking he was leaving town under circumstances implying Gates had "few, if any, friends," and no one would come to his defense. They added, "When public sentiment is purer than its organs it will reform or destroy them.—So will it be in the institution of slavery, or any other that violates human nature."[94]

Many more must have listened as well; Republicans decisively won their entire ticket of state and county officers in Jefferson County's election of August 4, 1856. As W. W. Junkin proclaimed,

> With pride we herald the glad tidings to the State, and ... throughout the country, that Old Jefferson has, by the largest majority ever given in the county, endorsed throughout the glorious principles of the Republican party.—FREEDOM stands triumphantly erect in our noble county, and the routed forces of the slavery extensionists. The Administration forces fought with desperation, but Freedom's army swept the field, leaving not a vestige of hunkerism to mark the spot where waved the black banner of slavery extensionists.[95]

The Kansas bloodshed continued; on September 3, Rev. L. B. Dennis wrote from Lawrence directly to his friend Jesse Byrkit, a Fairfield UGRR station master, commiserating over the "godless" denials of Fairfield's pro-slavery "dough-faces" and describing first-hand Kansas's horrific state of war. In the past three weeks, men had been scalped, women ravished, horses stolen, houses burned, and Free-Staters evicted in pro-slavery Border Ruffians' attacks in Lawrence, Osawatomie, and Leavenworth, all abetted by the pro-slavery Administration. W. W. Junkin published long excerpts of his letter in the *Fair-*

---

[94] *Fairfield Ledger*, July 31, 1856, p. 2, col. 5.
[95] *Fairfield Ledger*, Aug. 7, 1856, p. 2, col. 1.

field Ledger. Junkin's next three columns carried the letter of "John Smith," also written September 3 from Lawrence, Kansas, to the "*St. Louis Democrat*"—i.e., the anti-slavery newspaper *Missouri Democrat*, edited by U. S. Grant's longtime friend, George W. Fishback—describing the pro-slavery Senator Atchison's propaganda, the outrages of the Border Ruffians, and their battles with John Brown at South Middle Creek and Osawatomie, followed by a one-sided "battle" of 76 Border Ruffians against "five men, three boys and a woman" in a log house at Prairie City. Leavenworth also had bad news: "Murder and bloodshed there as well as here." The identical datelines and similar content of the two letters lead us to suspect "John Smith" may have been a *nom de guerre* of Rev. Levin B. Dennis.[96]

Many settlers were traveling through Fairfield to Kansas at this time, but in mid-September 1856, an armed wagon train of Free-Soil Kansas immigrant families from New England passing through Fairfield especially excited Democratic *Sentinel* editor David Sheward. He was President *pro tem* of Fairfield's Buchanan Club, which protested on the 18th to Iowa's Free-Soil and Republican Governor James W. Grimes, who had reportedly allowed the immigrants to take 1500 muskets from the Iowa State Arsenal on their way through Iowa City. With "indignation" the Club formally asserted their opinion that Governor Grimes "by thus countenancing the migration of *armed* men through the State, whose apparent purpose is the *invasion* of a sister province, is ... acting in violation of his duties as a Governor." They further resolved that "it is the duty of *all* good citizens to remonstrate and request him to interpose his authority to prevent the passage of those armed bodies of men through our State or resign his office." They also requested Democratic papers to publish their expressions, and Democratic clubs "to take action thereon."[97]

Grimes had conveniently left the arsenal key on his desk, where it was "borrowed" by the abolitionist journalist Richard Josiah Hinton, a Chartist weaver from Lancashire who had come to America from Eng-

---

[96] *Fairfield Ledger*, Sept. 18, 1856, p. 2, cols. 4-7.

[97] Charles J. Fulton, "Jefferson County Politics Before the Civil War," *Annals of Iowa*, vol. 11, no. 6 (July, 1914), pp. 441-442, http://ir.uiowa.edu/annals-of-iowa/vol11/iss6/11/. W. G. Wilson was secretary of Fairfield's Buchanan Club. The club's president *pro tem* was David Sheward of Ohio, whose Fairfield household (214-224) appeared next below that of fellow-printer William W. Junkin (213-223) in the 1856 state census. Between 1852 and 1857, David and his brother William H. Sheward published the *Iowa Sentinel*.

land in 1851 and was now going to Lawrence, Kansas. A correspondent for the Boston *Traveller* and the Chicago *Tribune*, Hinton would soon meet John Brown, and long afterwards write an admiring biography of him.[98]

The questionable legality of Grimes' action aside, the irony of Fairfield's Buchanan Club's protesting the Free-Soilers' "invasion" of Kansas was not lost on W. W. Junkin, who responded tartly:

> We suppose that the resolutions were passed to meet the case of the emigrant train which passed through our town during the past week. That train was made up of persons going to settle in good faith. They were well supplied with everything to make their new homes comfortable, and were also provided with weapons to defend their homes from the ravages of the pro-slavery desperadoes who support themselves by robbing and murdering free State settlers. It is this last item which roused the ire of the pro-slavery club of this place. They do not like the idea of free State men being prepared for self-defense – such preparation will prevent Buchanan's Missouri supporters from making their living by robbing the settlers of the Territory, and we suppose it is for this reason that they want Governor Grimes to either stop these emigrants or resign his office. Many of the men who passed thro' this place on their way to Kansas, had their wives and children with them, which shows beyond all controversy that the object of the persons comprising the train was to become actual settlers.[99]

On October 4, 1856, Jefferson County Democrats nominated their veteran statesman Col. William G. Coop for delegate to Iowa's upcoming Constitutional Convention, but recommended no changes to the constitution. On October 18, the Republicans nominated James F. Wilson for delegate, and presented a constitutional platform much like that suggested by James W. Grimes in Fairfield two and a half years earlier: against slavery extension and for establishing banks, popular election

---

[98] Richard Josiah Hinton, *John Brown and His Men: With Some Account of the Roads they Traveled to Reach Harper's Ferry*, New York: Funk & Wagnalls, 1894, pp. 55-56; Benjamin F. Gue, *History of Iowa: From the Earliest Times to the Beginning of the Twentieth Century,* Vol. I, New York: The Century History Co., 1903, p. 375. After Hinton and his associates brought the guns to Kansas, Reverend Pardee Butler transferred them to the Free-State commanders. Reverend Butler had moved in 1855 from Cedar County, Iowa, to Kansas, where by the following year he had several times been attacked by pro-slavery mobs.

[99] *Fairfield Ledger*, September 25, 1856, p. 2, col. 4.

of Supreme Court judges, and a more convenient general election date, together with fair salaries for state and county officers, and popular vote on state expenditures. In the elections of November 4, 1856, James F. Wilson won narrowly, by 1207 to 1122 votes, over W. G. Coop, and was a delegate to the Constitutional Convention. Article I, Section 1 of Iowa's new constitution would read:

> *All men are, by nature, free and equal*, and have certain inalienable rights—among which are those of enjoying and defending life and liberty, acquiring, possessing and protecting property, and pursuing and obtaining safety and happiness. [Italics added.]

In November 1856, Jefferson County also chose the Republican J. C. Frémont for President over the Democratic James Buchanan by 1188 votes to 1023, with 206 votes for the Know Nothing or American Party's Millard Fillmore. The State of Iowa went Republican as well, but the country as a whole voted for Buchanan and another four years of a pro-slavery, Southern-sympathizing administration which would continue to uphold the federal law against helping fugitives escape from captivity.

1856-1857: Fairfield on the "Direct Line" of the UGRR

By September 1856, the Quaker UGRR operators Benjamin D. and Rachel (Coppock) Pierce moved to Fairfield from Newton, Iowa, where Benjamin had stood firmly with the abolitionists and had spoken out against slavery in some 40 counties of Iowa, "many times at the peril of his life." A carpenter and a farmer, Benjamin also "took an active part" in helping Black freedom-seekers "by means of the celebrated Underground Railroad." Newton was midway between the UGRR stations of Mitchellville and Grinnell, about 20 miles from each, and was thus on the UGRR trail soon made famous by John Brown, several of whose "soldiers," Edwin and Barclay Coppoc, were Rachel Pierce's cousins. But Benjamin Pierce's biographer says Newton was then "off the direct line," and Benjamin and Rachel moved to Fairfield to "better assist in that work," thus implying both that the Pierces moved to Fairfield specifically to work on the UGRR, and that Fairfield was already on "the direct line" as a significant UGRR nexus. The Pierces soon became prominent UGRR station masters in Fairfield, where Rachel was also an accomplished midwife.[100]

Also coming to Fairfield at this time were Fairfield Female Seminary teachers Sue and Kate McBeth, who had just kept a UGRR station in Ohio, and they may well have continued their service here.[101]

In late 1856 or very early 1857, soon after the arrival of the Pierces and McBeths, freed Blacks Nancy (Hale) and James H. Yancey moved to Fairfield from Ohio with their children, the first African-American family to settle here permanently. James established a barber shop and after the Civil War spoke around the state for Blacks' and women's suffrage. Nancy was Oberlin-educated. As a Black female, this was doubly rare and almost unheard-of at the time. She reportedly "spent much time" teaching, but chose to earn more and stay with her family by starting a laundry business, which she ran successfully for many years. The Yanceys were shining exemplars of culture and education, and were also prominent UGRR station masters in antebellum Fairfield.

1857: Dred Scott: No Rights Which the White Man was Bound to Respect

Meanwhile, Blacks were not faring so well in Missouri, nor in the U. S. Supreme Court. After working its way through the legal system for ten years, the Dred Scott case was now a political rallying-point. Beautifully summing up the South's hardened attitude toward slavery was Supreme Court Chief Justice Roger Taney, who in his 1857 Dred Scott decision, ruled that a slave-holder had a right to keep his slaves in a free territory, and infamously said:

> [Blacks] had for more than a century before been regarded as beings of an inferior order, and altogether unfit to associate with the white race, either in social or political relations; and so far inferior, that they

---

[100] *Portrait and Biographical Album of Jefferson and Van Buren Counties, Iowa...*, Chicago: Lake City Publishing Co., 1890, p. 303.

[101] For the McBeths as UGRR stationmasters, see Rory Goff, *Who's Who in the Anti-Slavery and Underground Railroad Networks in Fairfield, Iowa*, Fairfield, Iowa: Merrymeeting Archives, 2018; Mary Ellen Snodgrass, *Settlers of the American West: The Lives of 231 Notable Pioneers*, Jefferson, North Carolina: McFarland & Co., Inc., 2015, pp. 107-109; Allen Conrad Morrill and Eleanor Dunlap Morrill, *Out of the Blanket: The Story of Sue and Kate McBeth, Missionaries to the Nez Perces*, Moscow, Idaho: The University Press of Idaho, 1978, passim.

had no rights which the white man was bound to respect; and that the negro might justly and lawfully be reduced to slavery for his benefit. He was bought and sold, and treated as an ordinary article of merchandise and traffic, whenever a profit could be made by it.

Fairfield Ledger editor William Wallace Junkin immediately printed an article from the Albany *Evening Journal* pointing out that the same men ratifying the Constitution had only two months earlier ratified the Ordinance of 1787 prohibiting slavery in all U. S. territories. Junkin added the Dred Scott decision was "contemptible," because it overruled all previous decisions on slavery and defied the construction of the Constitution given by its own framers.

> It goes farther, and endeavors to turn this country into a mere machine for the protection and perpetuation of human slavery, and in effect proclaims that slavery is the general rule, and Freedom the exception.... It is the step which brings the plantation aristocrat to the very doors of the free States, with his miserable bondmen at his heels, and bids him knock, that ... he be permitted to desecrate the free soil of the free States with the feet of his slaves.

W. W. Junkin appended the Detroit *Advertiser*'s biographical sketch calling Taney "a partizan of such bitterness" that the Senate had rightly rejected his appointment twice, having no confidence in his honesty. "He will soon resign, it is said, to give Mr. Buchanan a chance to appoint. Good riddance." Junkin concluded he was cheered by the people's "almost universal" disapprobation. The judges' dissenting minority opinions and conflicting majority views "badly shattered" the decision's legal effect and proved it could not stand long. It was too transparently a partisan "straw thrown out to the rapidly sinking pro-slavery Democracy," a "batch of rotten contradictions, fitted better for the realms of barbarism" and Judge Jeffries' infamous medieval Star Chamber "than for the Republic of the United States."[102]

In another *Ledger* column, the ardent abolitionist Samuel A. Robb of Brookville responded to an article in Fairfield's pro-slavery *Sentinel* which claimed that African Blacks were cruelly enslaving and killing each other, and that slavery in the "Southern States" was "a benign, nay, a heavenly institution, and our Southern brethren deserve the thanks of the whole Christian world" for having "ameliorated, in such

---

[102] *Fairfield Ledger*, March 26, 1857, p. 2, col. 2.

striking contrast" with their African brethren, "three and a half millions of negroes." Robb pointed out that Europeans had created the African slave market, that the Czar of Russia had cruelly enslaved millions of his own slaves and no one was "trying to ameliorate their condition," and that "these negro ameliorators infuse a considerable amount of their own blood into the African and then whip and sell it." Robb concluded by asking:

> Wherein is the condition of the African ameliorated in this country? Are they better fed according to their labor? or better clad according to climate? Are they not prohibited from learning to read and write? Are they not prohibited from meeting together to worship their God? In short, what better is their condition than the horse, and what worse is it in Africa?
>
> But, admit that it has bettered the condition of some Africans, does not the world's history teach us that involuntary labor degrades voluntary labor, and where labor becomes disgraced the nation becomes debased, and no nation can then long flourish? [103]

W. W. Junkin criticized the Dred Scott decision nearly every week thereafter for months. The decision still rankled more than a year later, when Junkin sarcastically quoted Taney when lambasting the "fanfaronading" or arrogant boasting of young Fairfielders who attempted to intimidate a Black barber in town:

> About twenty of the "young bloods" of our city, who have an eye single to the moral welfare of this community, organized themselves into a committee of one for the purpose of notifying James Lucas, a mulatto barber, that it wouldn't be good for his health to be found here after twenty-four hours. Lucas has been here during a part of the winter taking care of his brother's family, and it was supposed, suspicioned or thought that, because he lodged at a neighbor's house, there was "something rotten in Denmark." Hence the outbreak of virtuous indignation. We also suppose that they went on the principle that "a nigger has no rights which a white man (or boys) is bound to respect."[104]

---

[103] *Fairfield Ledger*, March 26, 1857, p. 2, col. 4.

[104] *Fairfield Ledger*, July 29, 1858, p. 3, col. 1. Junkin concluded, "Lucas did not leave as requested, and we suppose the valiant company have disbanded and returned their arms, and that the spokesman for the evening is trying to jump through another window. 'Fuss and feathers.'"

And indeed, Taney's attitude disgusted much of the North. While most Yankees were at least somewhat racist themselves, the majority actively disliked what they saw as the boastful, arrogant tyranny of slave-holding. They still generally preferred to conciliate rather than confront the South and risk war or secession, but the demographics were slowly shifting.

1857-58: Republican Growth in Jefferson County

Fairfield Township's Republicans met in June 1857 and with A. H. Brown as chairman and W. M. Clark as secretary, selected Philander Chandler, R. R. Mills, D. C. Brown, UGRR operator J. A. McKemey, W. M. Read, E. E. Easton, Kirby Caviness, Thos. Moorman, George Stever, H. B. Mitchell, C. W. Slagle, J. A. Cunningham, and W. M. Clark as delegates to the County Convention of June 27. The County Convention chose W. L. S. Simmons of Locust Grove as their candidate for county judge, J. A. McKemey of Fairfield for recorder and treasurer, W. D. Clapp of Lockridge for sheriff, A. R. Fulton of Fairfield for surveyor, and G. W. Pancoast of Liberty for coroner. Brookville abolitionist Samuel Robb wrote to the *Ledger* affirming the stellar character of W. L. S. Simmons, whom he had known "from infancy" in Clermont County, Ohio, and who was self-made. He was a Freesoiler, former Whig, and Republican for Fremont, who "like a *considerable* number of us... believes that *slavery* was the downfall of the Republics of Greece and Rome, and very powerfully threatens the downfall of *this*." Junkin encouraged Republicans to work for their principles, remembering that "every candidate on the ticket of the Sham-Democracy"—local, state, and national—"endorses all of the ultra-pro-Slavery doctrines of that party—the Nebraska bill, Cincinnati platform, Buchanan's Inaugural and the Dred Scott decision." But his words were as yet in vain; Democrats won all the offices that August except for surveyor, won by A. R. Fulton.[105]

---

[105] *Fairfield Ledger*: June 25, 1857, p. 2, col. 4; July 9, 1857, cols. 1, 5, 4 For 1857, Thomas McCullough was re-elected county judge; George Shriner was sheriff; Samuel Clinton was prosecuting attorney; Thomas B. Shamp was re-elected treasurer and recorder; A. R. Fulton was re-elected surveyor, and J. M. Wright was re-elected coroner. Charles J. Fulton, *History of Jefferson County, Iowa: A Record of Settlement,*

More work clearly needed to be done. Perhaps in part because of that egregious Dred Scott decision, when the Jefferson County Republicans met in convention at the Fairfield court-house on September 19, 1857, many new anti-slavery recruits reinforced the old guard. On motion of C. W. Slagle, J. J. Cowen of Des Moines Township was chosen president, while A. R. Fulton was again secretary. The township delegates had new faces among the old, as James Beatty, R. R. Mills, D. C. Brown, A. R. Fulton, P. S. Patton, Kirby Caviness, Charles Coleman, C. W. Slagle, Mathew Clark, A. Hemphill, James McFee, and Milton Crail appeared for Fairfield; J. Cochran, W. S. Lynch, John E. Dunham, and L. D. Parker for Round Prairie, Abner Frazier, Isaac Ellis, A. "Emery" [Emry], T. E. Osborn, Solomon Nordyke, Alexander Ireland, and Isaac Garmoe for Penn; Richard Gaines, W. Hutchin, and John L. Hadley for Black Hawk; O. O. Sheldon and F. T. Humphrey for Cedar; Thomas Byers, Joseph Holsinger, G. P. Loomis, George W. Pancoast, and S. H. Merritt for Liberty; M. Henderson and J. Bardine for Locust Grove; Jacob Ramey and George W. Sisson for Polk; Joseph Junkin, Reuben Chilcott, J. M. Strong, Waltus Collins, Archer Green, and John Hoagland for Lockridge; and Samuel Eshelman and J. J. Cowan for Des Moines.

For state representatives the convention nominated J. F. Wilson and Thomas Moorman, the other candidates being Mathew Clark, Richard Gaines, J. H. Allender, Isaac Ellis, W. W. Cottle, H. B. Mitchell, J. J. Cowan, A. R. Fulton, and Isaac Garmoe. Both Republican candidates won and served in 1857-59; James F. Wilson would then serve as state senator in 1859-61 and U. S. Congressman in 1861-69, where he would continue to be a driving force for Blacks' civil rights.

The 1857 County Convention chose UGRR operator Joseph A. McKemey as the Republican nominee for recorder and treasurer by acclamation. McKemey won, and would be re-elected annually until 1863. The convention also endorsed the Republican platform of 1856, condemned the recent pro-slavery Dred Scott decision, congratulated Iowans on their new constitution, endorsed Governor Grimes's administration, and recommended adopting a state banking system.[106]

More new faces joined the active Republican ranks in 1858: At a Republican convention in Fairfield that March, veteran John McCleary

---

Organization, Progress and Achievement, Chicago: S. J. Clarke Publishing Co., 1912-1914, Vol. 1, p. 415.

[106] *Fairfield Ledger*, Sept. 24, 1857, p. 1, col. 6.

chaired, but J. H. Beatty was secretary at the nominations of a township ticket, consisting of T. D. Evans, P. S. Patton, and G. W. Stewart for justices; James Cunningham, James Graham, and John P. Ramsay for constables; Thomas Moorman, J. M. Slagle, and UGRR station master Jesse Byrkit for trustees; and William Long for clerk. In Pleasant Plain on September 9, James Harvey presided, with I. W. Ellis as secretary, at a Republican convention which passed powerful resolutions against slavery, and condemned President Buchanan for trying to force slavery on Kansas.

> Whereas, We believe that slavery, as it exists in these United States, is at war with the Christian Religion, at variance with the Declaration of Independence, and subversive of the peace, harmony, and prosperity of the whole country. Therefore
>
> *Resolved*, That we look upon the efforts of the present administration to extend, strengthen and perpetuate this wicked and dangerous institution, with fearful forebodings for the peace, prosperity and permanency of our government.
>
> *Resolved*, That the efforts which have been made by the present administration to force slavery upon the people of Kansas in opposition to the well known wishes of a large majority of her citizens, manifest a spirit of despotism unparalleled in the history of our country.
>
> *Resolved*, That of all northern men who have ever filled the Presidential Chair no one has ever so meanly, basely or treacherously submitted to southern dictation, as James Buchanan, our present Chief Magistrate.
>
> *Resolved*, That in the election of persons to offices, the interests of the people, alone, should be regarded, and personal interests entirely disregarded.[107]

Jefferson County's Republicans convened in Fairfield on September 11, 1858 to nominate a candidate for district court clerk. W. M. Read called the convention to order and moved that Thomas Byres, Esq. be called to the chair, and J. B. McLean appointed secretary. W. M. Read, C. W. Slagle, W. McLean, Thomas Moorman, W. M. Clark, R. R. Mills, A. H. Brown, A. M. Scott, Mathew Clark, James Williamson, James "Krail" [Crail], and D. J. Evans were the delegates for Fairfield Township; J. B. McLean, Samuel Maritt, Gilbert P. Loomis, George W. Pancoast, Thomas Byres, and J. F. Robb for Liberty Township; Samuel Eshelman, James Cowan, Jacob "L." Lamb [UGRR oper-

---

[107] *Fairfield Ledger*, Sept. 16, 1858, p. 2, col. 5.

ator Jacob T. Lamb], and S. Pumphrey for Des Moines Township; K. Caviness, James Beatty, E. Hoops, and G. Frush for Buchanan; W. S. Lynch, John Houston, S. W. Taylor, and J. McLoney for Round Prairie; T. F. Humphrey and Andrew Yount for Cedar; I. G. Rhodes, John Pheasant, and H. Gorsuch for Walnut; I. "Hutchens" [Hutchin], T. A. Robb, and Richard Gaines for Black Hawk; Isaac Ellis, Christian Ossart, John Williams, William L. Koontz, UGRR operator C. O. Stanton, and John Ecroyd for Penn; Mr. Robinson for Polk; and Joseph Bartine, Samuel Robb, and Joseph Ball for Locust Grove.

A. M. Scott and Isaac Hutchins were appointed tellers; R. F. Ratcliff emerged as nominee for re-election for district court clerk with 34 votes over A. R. Fulton (7 votes) W. Long (6 votes), and S. Robb (1 vote). The convention adopted C. W. Slagle's resolution that the taxpayers thank the Republican State Board of Equalization for reducing the state tax from two to one and one-half mills on the dollar, and for reducing Iowans' aggregate tax by $100,000 from last year.[108]

W. W. Junkin called for still more Republicans to awaken the county to the injustices of the pro-slavery Democrats:

> We hope our Republican friends ... will now put themselves in the harness and go to work with a determination to carry Jefferson County by a larger majority than ever before.... The iniquities of the present administration are sufficient to sink the money-squandering, pro-slavery democracy deeper than they have ever been sunk in this county before. Every Republican should do his share of the work in bringing these iniquities fairly before the minds of the people. Go to work then, every man, and let us... do our duty in true Republican style.[109]

Each Republican evidently did do "his share" in bringing the pro-slavery Democracy's "iniquities fairly before the minds of the people." Jefferson County's voters elected Robert F. Ratcliff as their district court clerk.

1859-1860: Christian S. Byrkit on the UGRR

In *A Derailment on the Railway Invisible*, Christian S. Byrkit says he "occasionally" helped his parents as a UGRR conductor "as the war

---

[108] *Fairfield Ledger*, Sept. 16, 1858, p. 2, cols. 4-5.
[109] Ibid., p. 2, col. 2.

clouds thickened and neighbors grew suspicious of each other," probably in the year or two before the Civil War when he was ten or eleven years old; he turned twelve almost two months into the war. Men apparently did most UGRR conducting, and women and children did more UGRR station-work in feeding, clothing, and caring for freedom-seekers, but women and children did sometimes conduct as well, dangerous though that job was.

Christian did not say whether he was his family's first or only conductor, but he probably was not. His older brother Archibald had apprenticed and lived with the alleged UGRR operator Daniel Mendenhall—whose wife Susannah (Pierce) was a niece of Fairfield's UGRR station master Benjamin Pierce—as early as 1854, when Archie was 18 and Christian was only five. Perhaps Archie, like Christian, was conducting at the age of 10 or 11, around 1846-47. Archie had two older brothers who may have started even earlier.

Born in Fairfield on June 4, 1849, Christian was the youngest of nine children of Isabella (Woods) of Kentucky and Jesse Hoover Byrkit, a carpenter and wagon-maker from Asheville, North Carolina. Jesse was born a Quaker; Isabella was a Presbyterian minister's daughter, and in Fairfield they brought their children up in the Methodist Church. Two of Chris's brothers became Methodist ministers. The Byrkits came to Fairfield in 1843-44, and lived at the southeast corner of East Burlington and B Street, at what is now 200 East Burlington Avenue. The Byrkits may have served on the UGRR in Fairfield as soon as they moved here, but they were certainly participating by around 1859.

1859: Republicans Consolidate

On June 16, 1859, Jefferson County's Republicans met at Fairfield's court house and called Dr. W. W. Cottle to the chair and appointed W. W. Junkin secretary. J. F. Wilson, Dr. C. S. Clarke, and R. F. Ratcliff comprised a committee selecting delegates to the State Convention on June 22 in Des Moines: C. W. Slagle, UGRR operator J. A. McKemey, W. M. Read, UGRR operator Richard Gaines, George C. Fry, M. Green, John Cochran, John Spielman, Joseph Junkin, J. F. Wilson, and UGRR operator C. O. Stanton.[110]

---

[110] *Fairfield Ledger*, June 17, 1859, p. 2, col. 7.

Another Republican County Convention on August 27, 1859 called William Long to the chair and, on W. M. Clark's motion, appointed T. A. Robb secretary, *pro tem.* On Dr. C. S. Clarke's motion, Long appointed Nathan Birkhimer, Alexander Fulton, and John H. Hill to examine and report delegates' credentials; C. S. Clarke, Isaac W. Ellis, Reuben Chilcott, S. C. Farmer, and Joseph Ball to report the names of the convention's permanent officers; and W. Z. Hobson, Charles Reed, and George Howell to report resolutions and name a county central committee. W. M. Read of Fairfield was named president; G. C. Fry of Liberty and Archer Green of Lockridge were vice presidents; J. H. Beatty of Fairfield and W. Z. Hobson of Black Hawk were secretaries.

In the afternoon, the Republicans appointed A. R. Fulton as assistant secretary, and named the delegates: Dr. G. A. Smith, Peter Slimmer, W. F. Dustin and G. C. Fry from Liberty; G. W. Calfee, John Winsell, W. H. Copeland, and James Cowen from Des Moines; Joseph Fell and Morgan Flower from Cedar; Isaac Galliher, E. J. Gilliam, James McCullough, and Jacob Metz from Round Prairie; Reuben Chilcott, Archer Green, and David Rock from Lockridge; Samuel C. Farmer, Nathan Birkhimer and William Cummings from Buchanan; Dr. C. S. Clarke, J. B. Horn, Mungo Ramsay, J. H. Beatty, Dr. Charles Reed, Alexander Fulton, W. M. Read, John H. Hill, John Williamson, A. P. Heaton, Milton Crail, George Howell, William "Hoaglin" [Hoagland], and Marmaduke Green from Fairfield; Joseph Ball and R. "Nimicks" [Nimocks] from Locust Grove; George Frush, Jacob Ramey, "William" [Wesley] B. Campbell, and John Jones from Polk Township; W. Z. Hobson and T. A. Robb from Black Hawk; I. W. Ellis, Isaac Garmoe, Thomas Hodson, J. F. Hodson, Thomas "Talbott" [Talbert] and Elliott Hiatt from Penn; and C. W. Wood, Nathan Cole and Daniel Creegan from Walnut.

With W. H. Copeland and C. S. Clarke as tellers, the convention nominated J. F. Wilson for state senator and W. W. Cottle and Matthew Clark for representatives, Jacob H. Allender for county judge over fellow-Republican candidates S. P. Majors and S. M. Bickford, and J. A. McKemey again for treasurer and recorder. They also nominated J. F. Robb for sheriff over candidates I. H. Brown, J. A. Cunningham, Joseph Junkin, W. M. Campbell, H. W. Hutchin, J. M. Grafton, John Woods, B. B. Tuttle, and W. D. Clapp. They selected R. S. Hughes for school superintendent over candidates A. M. Scott, Rev. Andrew Axline, and Dr. J. M. Shaffer. They nominated A. R. Fulton for surveyor, and Thomas Barnes of Fairfield for coroner.[111]

The Republicans virtually swept the election, winning all of their offices except that of judge, who was "defeated by about 19 majority." As Junkin exulted in huge headlines, "Democracy Down in the Mouth! Railroad Company Annihilated! Republicans Have Given a Majority for Their State Ticket and Elected Their Senator, Representatives, Treasurer, Sheriff, Superintendent, Surveyor and Coroner. Majorities Ranging from Sixty to One Hundred and Twenty." Republicans also swept the township ticket, where John Mount, William Long, and W. M. Moore were elected trustees; J. A. Cunningham, Peter Gow, and Thomas Barnes were elected constables; W. S. Moore was elected clerk, and B. H. "Canady" [Canaday] won the race for assessor.[112]

Jan. 1860: Halting the Reverse Underground Railroad in Fairfield

On one occasion before the Civil War, anti-slavery Fairfielders believed they saw the Reverse Underground Railroad in action in their town, and took steps to stop it. A Fairfield man signing himself "Union"—perhaps the *Ledger*'s editor, W. W. Junkin—described the "kidnapping" in a letter to the *Burlington Hawk-Eye* on January 31, 1860:

> John L. Curtis and James B. Little left Iowa City on Saturday evening, 29th inst., about dusk or, shortly after, in a two-horse carriage containing two negro girls, one of the age of 14, the other 13. They travelled all night, and early on Sunday morning called at Mr. Allen's, four miles south of Washington, for breakfast, &c. The fact that the party had travelled all night, and other circumstances, aroused the suspicion that all was not right, and influenced three gentlemen to follow them to Fairfield, where they obtained a warrant, which was placed in the hands of deputy Sheriff Cunningham, who with the gentlemen referred to, followed Messrs. Curtis and Little to Iowaville, where they were arrested after a scuffle and the handling of a loaded pistol. They were brought here to-day for examination before Justice Evans, but for want of evidence from Iowa City, on the part of the defense, the examination is set for next Saturday, 4th of February, 10 o'clock, A.M.

---

[111] *Fairfield Ledger*, Sept. 2, 1859, p. 3, col. 2.
[112] *Fairfield Ledger*, Oct. 14, 1859, p. 2, cols. 1, 4.

> This is the first case of the kind that has ever occurred in our city. There is no undue excitement; the general feeling is to see that justice is had in the case. Mr. Curtis claims that the girls were the property of his wife in Tennessee; that he moved to Iowa four years ago; that the girls were sent to Iowa City two years ago; that he is their guardian; that he has their indentures; that he was going to Missouri to purchase a farm; that he was taking the girls to Missouri to see their mother, &c. &c. They seem very much attached to "Master John," and have entire confidence in him. As to what Master John's intentions were, and how the case will terminate, will be better known after the examination. Mr. Curtis gave bonds for $400. Mr. Little, for want of bonds, was committed.
>
> I am just informed that the Sheriff from Johnson county is here after Mr. Curtis. If so the train on the "overground" Railroad will be stopped.

The writer added that Messrs. Stubbs, Slagle & Acheson would be lawyers for the State, with C. Negus for the defense. Doubtless the attorneys were all satisfied with their various roles; Ex-Quaker Daniel P. Stubbs, Christian W. Slagle and George Acheson were all notably anti-slavery, while Charles Negus was pro-slavery and had married into a Virginian family of slave-holders.

"Union" then appended another letter with "More about the Kidnapping Case," confirming that the Johnson County sheriff was in Fairfield with a warrant for John L. Curtis, and adding:

> Previous to the arrival of the Sheriff there was "more or less" sympathy for Curtis. It was even said that he was a high-minded honorable Southern gentleman who had a natural and legal right to the *persons* of *his* wards, he being their guardian, &c. It had come to a h—l of a pass that a gentleman could not even look at a nigger without the *risk* of an arrest, but *after* the Sheriff came from Iowa City it was talked about that Curtis was a d——d Yankee adventurer who evidently intended to make a good thing out of the negro children, *provided* he was not molested on his way to Missouri. His generous bondsmen gave him up at once to the tender mercies of our Sheriff, who kindly showed him to bed in our jail. To-day he has been handed over to the Sheriff of Johnson County, who left here for Iowa City about noon, and I refer you in that direction for further information as to the results of this arrest.
>
> I learn that the little girls were indentured to Curtis. He was bound to educate, clothe, feed and provide for them; that on the be-

lief that he had neglected his duty in whole or in part, the County Judge had a notice served on him last Friday to appear before him on Monday to show cause why he should not appoint another guardian, &c.

Whilst professing to prepare for that investigation on Monday, he quietly prepared to run the little girls into Missouri, and for that purpose left Iowa City about 11 o'clock on Saturday night, but through the vigilance of Messrs. Allen, Vincent and McCoy of Washington county, he was arrested.[113]

On February 4, Davenport's *Daily Gazette* reprinted an article from Iowa City's *Republican* to counter two letters in the *Democrat* giving a "scandalously one-sided account of the attempt to run two young negresses from that city into Missouri:"

About five years since, a Mr. J. L. Curtis came from Tennessee and settled in Pleasant Valley, in this county, bringing with him, to use a southern phrase, two "likely" negro children, then of the ages of five and seven years. Upon his own application, Curtis was appointed "Guardian" over these unfortunate descendants of Ham. Everything moved along quietly, until recently, when it began to be whispered about, and generally believed, that the girls were slaves, and that the guardianship of Curtis was only a ruse to hoodwink the unsuspecting Hawkeyes. Complaint was made by some of Curtis' neighbors, that he was not performing his duty as such Guardian, &c. Some months since, Curtis left his farm in the Valley, and moved to this city. On Saturday last, the feeling on the subject getting up to fever heat, Curtis concluded that "discretion was the better part of valor," and that night, between 11 and 12 o'clock, he vamoosed the ranche, taking with him the cause of the disturbance, leaving word that he was going to Lancaster, Mo., to leave the "chattels" with some "relatives" said to be living in that convenient locality, until the question should be finally settled. Information was filed, and a warrant was issued by his Honor, Judge Miller, for kidnapping, under the provisions of the code on this subject, and about twenty-four hours after Curtis' departure, officers were in pursuit. It is supposed, however, that the Missouri line would protect him before he could be overtaken. These are the facts as we have them from persons friendly to Curtis.

If these girls have never been liberated—which is the general belief in this city—the outrage of having slaves held in this city and

---

[113] *Burlington Daily Hawk-Eye*, Feb. 1, 1860, p. 2, col. 4.

State, contrary to the spirit of the law, and contrary to the feelings of every honorable instinct of the human heart, the fact should be ferreted out and the violator of the law punished to the last extremity of the same. The belief here is also very general that the negresses have been or will be sold, and the money thus obtained for human flesh and immortal souls be pocketed. We have endeavored, as stated before, to state the case fairly, with no intention to do Mr. Curtis, or any one concerned, injustice.

Curtis is said to be a man of some wealth, owning the farm upon which he lived and some city property, but of late is thought to be pretty "hard up."

The *Gazette* also included an article from the Democratic Fairfield newspaper *Jeffersonian*, edited by the pro-slavery Samuel Jacobs and H. N. Moore:

The Fairfield *Jeffersonian* of the 31st, contains the following, showing that Curtis was overtaken:

Arrested.—As we go to press, the Court House is filled to overflowing with people anxiously awaiting the decision of the Justice's Court in a supposed case of kidnapping, which seems to have elicited considerable attention, and caused no little excitement. The circumstances, as near as we can gather them, are as follows: Sunday afternoon about 4 o'clock, two men passed through town with two negroes in their carriage, on their way as was supposed, to Missouri,—About half an hour afterwards, some men arrived from Washington, in pursuit of them, claiming that they had stolen the negroes, and were taking them to Missouri to sell them into slavery.

The party here was joined by Deputy Sheriff Cunningham and Sheriff Robb, who pursued the supposed "nigger stealers," and succeeded in arresting them at Iowaville, in Van Buren county, on Monday morning. Monday about noon they arrived in this city with the prisoners, and the trial is now progressing at the Court House. The negroes are females—one about 16 years of age, and the other about 10 years old.

P. S.—Since the above was written, we understand the man has been held to bail in the sum of $400 for his appearance next Saturday.[114]

That modest, passive-tense "P. S." was probably the *Jeffersonian*'s downfall. On February 10, W. W. Junkin gleefully exposed the three

---

[114] *Davenport Daily Gazette*, Feb. 4, 1860, p. 2, cols. 3-4.

prominent, pro-slavery Fairfield Democrats who had praised the "gentlemanly" J. L. Curtis and posted his $400 bond, while snubbing his associate, J. B. Little:

> ... The leaders of that party took strong grounds in favor of the kidnapper Curtis, asserting that he was an innocent man, and that a great outrage had been perpetrated in arresting him while peaceably traveling through the State in company with two negro children going to see their mother. They also passed many encomiums upon the gentlemanly bearing of the much abused Curtis. One admirer went so far as to say, while Curtis was on trial, that his face was enough to clear him in any court of justice. Poor Little received little comfort, and was lodged in jail. Democratic sympathy ran so high, however, for the principal rogue, that three of the most prominent and philanthropic gentlemen of that party went security for his appearance at the trial on Saturday last. As these gentlemen were so benevolent, the presumption is that they will thank us for publishing their names for the benefit of the public generally.... The names of these gentlemen are known in this community as follows: Col JAMES THOMPSON, defeated candidate for Representative on the Democratic ticket last October. SAMUEL JACOBS, member of the Democratic County Central Committee, lately *one of the published* editors of the *Jeffersonian*, and now *the silent* editor of that sheet, the name of "H. N. Moore, Editor," to the contrary notwithstanding. WILLIAM L. HAMILTON, defeated candidate for Representative on the Democratic ticket in 1857.

Junkin added sarcastically,

> We take it for granted that they desire their "good deeds known of all men," else we might experience some diffidence in displaying their names in capital letters. What the action of these gentlemen would be in a case wherein a negro applied to them for assistance, our readers can judge. We believe they would make blood hounds of themselves for southern masters, and glory in returning a slave to bondage.[115]

---

[115] Junkin concluded, "In this connection we deem it due to our citizens that they all—with the exception, it may be, of a few extremists in both parties—desired to see justice done, and were satisfied that Curtis should have the negroes if he could prove that he was using no unlawful means to get them under his protection." In the next column was: "J. L. Curtis: ... this man has been bound over in the sum of $1000 for his appearance at the District Court in Johnson county to answer to the charge of kid-

Junkin added that Curtis was bound over for $1000 to appear in the Johnson County district court to answer to the charge of kidnapping, and that "Little has turned State's evidence and will make a clean breast of it," but at John L. Curtis's trial in Iowa City that spring, a jury of ten Democrats and five Republicans acquitted him. The census taken July 16, 1860 in Iowa City shows the two Black girls, Mary, 14, and Vessey, 10, still living in his household. We find no further record of the girls; John and Nancy later moved to Kankakee, Illinois, where he dealt first in horses and later in dry goods.[116]

The whole affair may have had more obvious repercussions in Fairfield than elsewhere. After Samuel Jacobs resigned as *Jeffersonian* editor, perhaps in shame, H. N. Moore continued the newspaper alone until closing it that fall, and soon afterwards, like a microcosm of Fairfield's larger shift from pro-slavery to anti-slavery sentiment, W. W. Junkin bought the material.[117]

---

napping. Little has turned State's evidence and will make a clean breast of it." *Fairfield Ledger*, Feb. 10, 1860, p. 2, cols. 1-2.

[116] Curtis's Iowa City trial is briefly covered in the *Fairfield Ledger*, April 27, 1860, p. 2, col. 1. John Lucky Curtiss was born in Cheshire, Conn. to Ruth M. and James H. Curtiss, a farmer prominent in town affairs. John went first to Shelbyville, Tenn., with his brother James H., who was in the dry goods business there. He went to Kankakee, Illinois, where he died leaving "one daughter and one son, James Anson Curtiss, b. 1846, now living in Meriden, Conn." Frederick Haines Curtiss, *A Genealogy of the Curtiss Family: Being a Record of the Descendants of Widow Elizabeth Curtiss...*, Boston: Rockwell and Churchill Press, 1903, p. 112. We find no other record of John Anson Curtiss. John L. Curtis had married Nancy R. Curtis on July 2, 1856 in Johnson Co., Iowa. The 1860 U. S. census for Iowa City shows John L. Curtis, 32, auctioneer with real estate of $5000 and personal estate of $250, b. in Conn.; his wife Nancy, 21, b. in Tenn.; Mattie, 1, b. in Iowa; with Mary, 14, Black; and Vessey, 10, Black, both b. in Tenn. On July 1, 1863, Civil War draft registration records show Curtis was a 36-yr-old grocer in Iowa City from Conn. The 1870 U. S. census for Kankakee, Kankakee Co., Illinois shows John L. Curtiss, 42, dealer in horses, real estate $1000, personal estate $3500, b. Conn.; Nancy R., 31, keeping house, b. Tenn.; and Mattie, 11, b. Iowa. The 1880 U.S. census for Kankakee shows J. L. Curtis, 55, b. Conn., dry-goods merchant with "Mary" R., 41, b. Tenn., keeping house, and Mattie, 21, b. Iowa. On Feb. 15, 1883, the *Kankakee Gazette* wrote, "John L. Curtis, an old resident and business man of Kankakee, has been compelled to announce that he will close out his business, owing to continued ill-health. He has been confined to the house for several weeks with a cancer on the lip." *Theakiki: A Quarterly Publication of Kankakee Valley Genealogical Society*, vol. 35, no. 2 (May, 2005), p. 10: https://tinyurl.com/theakiki.

[117] *The History of Jefferson County, Iowa, Containing a History of the County, its Cities, Towns, &c....*, Chicago: Western Historical Co., 1879, p. 482.

March 1860: The Irrepressible Republican Club of Fairfield

When Fairfield's Republican Club met in March, 1860, Chairman *pro tem* C. E. Noble called the meeting to order, and R. F. Ratcliff, C. S. Clarke, and "Finlay" [Finley] McMartin as the committee on constitution submitted that as Republicans they upheld the Declaration of Independence, the Federal Constitution, the rights of the states, and their union, and that the Constitution conferred upon Congress sovereign power over the territories; that "it is both the right and duty of Congress to prohibit in its Territories those twin relics of barbarism, Polygamy and Slavery;" while "neither Congress nor sister States, have any right to interfere with Slavery or any other institutions existing in any State." They heartily endorsed all the principles set forth in the recent call for the National Republican Convention at Chicago. They then submitted a constitution, with the first article naming their association the "Irrepressible Republican Club of Fairfield," and the other articles establishing an executive committee of president, vice president, secretary, treasurer, and corresponding secretary; a five-person finance committee to raise funds; a regular meeting time; and club membership by signing one's name to the constitution. On D. P. Stubb's motion, the club received the committee's report. The club then adopted William Long's motion to meet every Friday evening. Chairman Daniel Devecmon and J. A. McKemey and George Howell as the committee on permanent organization chose Samuel Mount, Sr., for president, Sumner M. Bickford for vice president, Wm. S. Moore for secretary, J. H. Beatty for corresponding secretary, and W. M. Clark for treasurer. On W. M. Clark's motion, they appointed W. M. Clark, J. F. Robb, J. A. McKemey, John Mount, and W. W. Junkin as a finance committee.[118]

June 1860: Fairfield's Wide Awakes for Lincoln: First in Iowa

Abraham Lincoln and Hannibal Hamlin emerged as the Republican Party's presidential and vice-presidential nominees at Chicago's Republican National Convention on May 16-18, 1860. Not quite a month later, on June 14, 1860, the Republicans of Fairfield met at the court-

---

[118] *Fairfield Ledger*, March 30, 1860, p. 2, cols. 3-4.

house to organize a company of Wide Awakes: part of a nationwide, grassroots, paramilitary movement of young Republicans, probably modeled on the old Wide Awake clubs of the Know Nothings and the more recent Wide Awake Societies of the Kansas Free-Staters. Fairfield's was the first Wide Awake club in Iowa.

On S. P. Majors' motion, the Republicans called Dr. C. S. Clarke to the chair. On R. F. Ratcliff's motion, they appointed W. S. Moore secretary, and on F. B. McConnell's motion, they resolved to name the company the Republican Wide Awakes of Fairfield. F. B. McConnell was appointed the company's financial agent.

The Wide Awakes invited all young men willing to endorse the sentiments of the Republican Party, and to abide by the association's rules. The officers would be a captain, a lieutenant, a treasurer, and a secretary. Each member had to equip himself with black cap, oilcloth cloak, and a swinging coal-oil torch, or pay $2 into the treasury for them. Each agreed to take part in torch-light processions during the presidential campaign, perform escort duty, attend night meetings and grand rallies, and act as a vigilance committee on election day. Each member was to share equally in expenses assessed with the association's consent and drawn by the captain's written order. He also pledged to refrain from all public swearing and noisy demonstrations, to obey his officers, and "comport himself in a decent and respectful manner."

James F. Wilson signed the Wide Awake "constitution" first, followed that night by Dr. Charles S. Clarke, S. P. Majors, John F. Robb, Robert F. Ratcliff, Daniel Devecmon, W. S. Moore, UGRR operator Joseph A. McKemey, William Long, M. Simmons, Philo Morehouse, Sumner M. Bickford, George Howell, John R. Shaffer, F. B. McConnell, L. P. Taylor, Thos. L. Pollard, John Mount, Samuel E. Bigelow, S. B. Woods, A. Turner, W. E. W. Sherfey, George A. Wells, Harry Jordan, L. T. Keck, Solomon Light, J. Thompson, William W. Junkin, and A. R. Fulton.

The association elected George A. Wells as captain, Dr. C. S. Clarke as lieutenant; A. R. Fulton as secretary, and Francis B. McConnell as treasurer. R. F. Ratcliff, George Howell, and William Long were appointed to procure a room to meet in. (The Wide-Awakes would rent a hall in William E. Sargent's building for drills and parades.) They adjourned to meet in the sheriff's office the next evening.[119]

That next night, June 15, thirty Wide Awakes held a torchlight procession, marching through some streets to the park, performing some drills, and calling for speeches. Hon. James F. Wilson urged them to "be ever on watch, and guard the purity of the ballot-box with unceasing vigilance;" C. W. Slagle was with the Wide Awakes all the time, and said they were "well prepared to meet the enemy." First, they had lights to light them out, second, they could burn them out, and if that wouldn't do, they could smoke them out. Mr. Scott gave a "very fine and effective little speech" to the Wide Awakes and a large audience. The company gave three hearty cheers for Lincoln and Hamlin, and then went to the court-house to deposit their uniform and arms.

At their meeting on June 20, Capt. Wells and Lt. Clarke resigned and were succeeded by Alvin Turner and George A. Wells, respectively, and J. "W." Shaffer [J. M. Shaffer] was elected 2nd Lieutenant. W. H. Sheward, John R. Shaffer, and A. R. Rusch were the club's musicians. During the campaign of 1860, Fairfield's Wide Awakes traveled 40 miles by train and 187 miles by wagon, marching in Glasgow, Birmingham, Libertyville, Agency City, Brookville, Washington, Salina, and Abingdon.

On June 16, 1860, the Republicans of Locust Grove Township organized a club; its president was John Gantz, with Joshua Wright as vice president, Samuel Robb secretary, J. Bardine as corresponding secretary, and Thomas Griffin as treasurer.[120]

The Republicans of Richland held a meeting on June 28, called to order by A. H. Smith, Esq., and they called B. A. Haycock, Esq., to the chair, and chose Mr. Plum, Esq., of Washington County as vice president, and W. Z. Hobson of Jefferson County as secretary. Isaac Hutchins moved they adopt their resolutions; they called J. F. Warner, Esq. to speak on the Homestead Bill; Hon. J. F. Wilson then spoke for two hours, and they adopted the resolution of M. B. Mills, Esq., of Washington County, "That there is, and can be, no such principle in God's Universe as the right of property in man." Esq. Plum, Judge Rockafellow, I. H. Garetson and M. Green gave short speeches, and H.

---

[119] *Fairfield Ledger*, June 22, 1860, p. 1, cols. 1-2.

[120] Charles J. Fulton, op. cit., vol. 1, pp. 310-311. For Ulysses S. Grant's drilling the Lincoln Wide-Awakes in Galena, see Ulysses S. Grant, *Personal Memoirs of U. S. Grant*, New York: Charles L. Webster & Company, 1885, vol. 1, p. 217.

W. Hutchins, Esq. moved that the proceedings be furnished to the Fairfield *Ledger* and the Sigourney *Republican.*

A Republican Club organized in Abingdon on July 7 with Cyrus McCracken as chairman *pro tem*, and a committee of Jacob Ramey, W. M. Campbell, and J. J. Sperry. They elected John H. Webb president, Cyrus McCracken vice president, W. M. Campbell secretary, and David Peters, treasurer.[121]

The "Lincoln Rangers," a company of seventy Republican horsemen from Brookville and Locust Grove, organized near the end of August, 1860, with J. A. Ireland as captain.

On September 1, six youths debated politics in Fairfield's park: Gilbert B. Kirkpatrick, I. N. Elliott, and George Strong offered the Republican side, while R. J. Mohr, A. G. Thompson, and W. A. Jones took the Democratic.

On September 22, 1860, a Republican County Convention met at Fairfield's court-house to nominate a candidate for district court clerk. B. B. Tuttle moved the meeting be called to order, by appointing Isaac "Hutchins" [Hutchin] chairman, and G. W. Pancoast, secretary. On M. Clark's motion, Chairman Hutchins appointed Samuel C. Farmer, M. Clark, and William Keech as a committee reporting the convention's permanent officers: John Hopkirk as chairman and G. W. Pancoast as secretary. A three-person credentials committee reported those entitled to seats at the convention: J. L. Hadley, I. Hutchins, and G. W. Sisson for Black Hawk; Silas Deeds and I. G. "Rhoads" [Rhodes] for Walnut; B. C. Andrews, UGRR operator C. O. Stanton, G. A. Hobson, E. K. Hobson, and J. W. Nicholson for Penn; G. Frush, J. H. Webb, and J. Ramey for Polk; James Beatty, S. C. Farmer, George Frush, and G. W. Vance for Buchanan; Joseph Bardine for Locust Grove; Archer Green, John Hopkirk, Nathaniel Simmons, and G. Anderson for Lockridge; W. G. Black and J. J. Cowen for Des Moines; none for Cedar; W. Keech, G. W. Pancoast, W. R. McCartney, and J. M. Bell for Liberty; and Columbus Hall, Henry Horton, John Strong, and John Horton for Round Prairie. S. P. Majors moved that a committee of three report the names of three people to serve on the central committee for the coming year. S. P. Majors, Jacob Ramey, and Joseph Bardine duly reported J. Cunningham, W. M. Clark, and W. W. Junkin as next year's central committee. With Wm. Z. Hobson, R. F. Ratcliff, Wm. Long, W. S. Moore, A. R. Fulton, and T. D. Evans as candidates, on W. W. Junkin's motion

---

[121] Charles J. Fulton, op. cit., vol. 1, p. 311.

after the fourth ballot, William Long was unanimously elected the Republican nominee for district court clerk. He would win and serve for four consecutive terms, 1861 to 1868. G. W. Pancoast moved and the convention resolved to publish its proceedings in the *Ledger*.[122]

Oct.-Nov. 1860: Fairfield's Grand Rally and Republican Victory

As the election neared, R. F. Ratcliff chaired the committee of arrangements for a mass Republican meeting and Wide Awake procession in Fairfield on October 17, 1860. A. R. Fulton was secretary of the committee, which included George Howell, James M. Slagle, UGRR operator J. A. McKemey, A. H. Brown, Daniel P. Stubbs, William Long, Daniel Devecmon, James A. Cunningham, John F. Robb, George W. Frush, Henry B. Mitchell, S. P. Majors, F. T. Humphrey, Robert S. Hughes, Wickliffe M. Clark, A. M. Scott, Marmaduke Green, James F. Wilson, and Alvin Turner. The committee selected Dr. Peter Walker as chief marshal, and Henry W. Hutchins, Jacob H. "Allinder" [Allender], Dr. S. W. Taylor, Wm. L. McLean, "Lorain" [Loren] Clark, Thomas D. Evans, Marmaduke Green, Alexander Fulton, C. W. Slagle, C. E. Noble, and John M. Woods as his assistants.

The rally's entertainment committee consisted of W. W. Junkin for the city's southeast quarter, R. F. Ratcliff for the southwest quarter, A. H. Brown for the northeast quarter, and William Long for the northwest quarter. The finance committee comprised Dr. C. S. Clarke, L. P. Taylor, and Daniel Young; the committee to examine and report on the procession route consisted of A. M. Scott, A. Turner, and A. H. Brown, and the committee of correspondence consisted of Jas. F. Wilson, C. W. Slagle, R. C. Brown, F. B. McConnell, C. S. Clarke, and D. P. Stubbs.[123]

The Republican meeting was the largest rally Fairfield had ever seen. Over 2,000 Wide Awakes, in 28 companies, came from New London, Mt. Pleasant, Salem, Rome, Eddyville, Ottumwa, Kirkville, Agency City, Bloomfield, Drakeville, Troy, Sigourney, Richland, Martinsburg, South English, Dutch Creek, Washington, Brighton, Richmond, Jackson, Keosauqua, Bentonsport, Bonaparte, Winchester, Birmingham, Glasgow, Pleasant Plain, and Fairfield.

---

[122] *Fairfield Ledger*, Sept. 28, 1860, p. 3, col. 2.
[123] *Fairfield Ledger*, Oct. 5, 1860, p. 2, col. 3.

Agency sent a company of white-dressed, blue-sashed "lady Wide Awakes" with spears flying "Lincoln and Hamlin" flags; other companies included the "Daughters of Abraham," over 100 "Minute Men of 1860," over 200 "Lincoln Guards" from Libertyville, in all about 600 riders, along with over 800 wagons and floats, making a procession five miles long. Three stands in Central Park hosted afternoon speeches by Governor Kirkwood, Senator Grimes, Senator Harlan, and others including A. M. Scott of Fairfield. That night, 2,200 Wide Awakes carried torches in parade, accented by skyrockets and Roman candles. The Democrats attempted to outdo the rally the following week, but were foiled by bad weather. [124]

Jefferson County cast 1,462 votes for Lincoln, 1,245 votes for Douglas and 38 votes for Bell. The steadfast heroes and heroines of anti-slavery and the UGRR had inspired the arousal of the Wide Awakes and the youth vote. Thanks to that heroism, in less than six years a despised and negligible third party with idealistic and progressive anti-slavery aspirations had swayed the hearts and minds of Jefferson County, Iowa, and of a majority of the United States, to give us one of our greatest presidents: Abraham Lincoln.

Feb. 7, 1861: Fairfield's Heartfelt Letter to the South

As the Southern states began to secede after Lincoln's election, U. S. Congressman John E. Bouligny of Louisiana strongly opposed his state's secession to join the Confederate States of America. On February 7, 1861, 77 Fairfielders wrote him a joint letter of thanks, lauding his courage and patriotism, and that of other Southern Unionists like Emerson Etheridge, Andrew Johnson, Andrew J. Clements, and Edmund J. (not Jefferson) Davis:

> Dear Sir—The undersigned, citizens of Iowa, having read your remarks in the House on February 5th ... would ask the liberty... to tender to you our thanks for your manly and bold declaration of devotion to the American Union.
>
> To speak bold words of patriotic devotion to our common country at this time, in your position...is a work of more than ordinary moral

---

[124] *Fairfield Ledger*, Oct. 19, 1860, p. 2, col. 2; Charles J. Fulton, op. cit., vol. 1, p. 313.

> courage, and cannot fail to strike fire in the hearts of all loyal American citizens in all sections of our beloved county.
>
> Here, in the remote West, you, sir, and your co-patriots have inspired us, and we swear anew our devotion to the Union and the Constitution *as it is.* Words are feeble to say how much we love and honor you.
>
> Here, in the free West, where there is no danger... of losing *caste*, or property or life, we feel that our declaration of devotion to the Union, compared with that of Bouligny, Etheridge, Andrew Johnson, Clemens, Davis, and others, is cheap and insignificant. They risk all, we nothing—they give evidence that they dare risk their liberty and lives for their country.

The Fairfielders lamented how most Southerners somehow saw the Republicans' simple Free-Soil position as aggressive abolitionism:

> Noble band of patriots! In the name of our country we thank you; aye, more, we reverence and ask God to bless you. *You* seem to understand us at the North, your people do not. We are their friends, but they think us enemies—Our principles are not understood at the South; we have been slandered and misrepresented, and these base slanders have been believed by your people. They will not hear us, and if they do they will not believe us.

They alluded to Republican willingness to permit slavery where it existed in the Southern states, but not to permit it to expand into the territories against the intent of the Constitution's signers:

> We trust and hope that Time and the unfolding of events will prove to them that under all circumstances and at all times we will respect and guard their rights in the Union and under the Constitution. As reasonable men they cannot demand more, as just men we cannot grant less, and as honorable men we cannot grant more.

They closed by hoping and trusting in eventual understanding and better times:

> We will, therefore, bide our time, until we are heard and understood, believing that then confidence, peace and good will, will be restored, and our happy but now distracted country will renew her course of honor and glory.
>
> You, and others that are with you, in devotion to the Union, will please accept our personal good wishes, and believe us,

Dear sir, yours respectfully and truly...[125]

The heartfelt "Northern" appeal was signed by Chas. S. Clarke, William Long, S. M. Bickford, A. M. Scott, M. Simmons, J. M. Whitham, D. Acheson, J. P. Roberds, John R. Shaffer, D. Beatty, S. B. Woods, L. F. Boerstler, R. K. Van Nostrand, B. Frank Ives, Jas. M. Slagle, E. L. Craine, Richard Bird, C. E. Noble, J. B. Rowland, George Craine, J. T. Rowland, W. B. Rowland, Thomas D. Evans, I. H. Brown, F. McMartin, D. Stanley, Geo. A. Wells, P. C. Arnold, S. Light, John Mount, D. P. Stubbs, Henry Eaton, Wm. R. Wells, Nathan Palmer, Robert F. Ratcliff, S. B. Whiting, Robert C. Brown, L. P. Taylor, Daniel Young, D. R. McCracken, James F. Crawford, J. M. Woods, Samuel Mount, Jr., G. W. Phelps, P. N. Woods, R. Wilkinson, E. C. Hampson, W. W. Junkin, J. Hamilton Beatty, Wm. H. Jordan, Wm. Foster, John F. Robb, W. C. Lewis, Charles W. Coleman, A. Case, John E. Cummings, Joseph Ball, Wickliffe M. Clark, George Howell, W. C. "Dias" [Dial], C. W. Slagle, Wm. Paine, George Acheson, Samuel Bigelow, H. G. Ross, H. H. McElderry, John E. Elbert, J. C. Snodgrass, John McCleery, Joseph Tillson, James F. Wilson, Ward Lamson, J. A. McKemey, Thos. L. Pollard, Daniel Devecmon, W. K. Alexander, and M. C. "Champ" [Shamp].

Nearly all Republicans, the signers never mentioned slavery, only their determination to uphold the South's Constitutional rights, granting no less and no more: a seemingly fair and reasonable position. But no prominent Democrats signed, except the "War Democrat" George Acheson who soon became a Republican, and the "patriot" Ward Lamson who was no longer allied to the land-office clique, and, surprisingly, photographer Moses C. Shamp, who soon appealed the freeing of the slave Ralph Robinson from him and David F. Philips at Shamp's house.

Feb. 15, 1861: 400 Jefferson County Republicans Petition Congress

Though the "Northern appeal" group letter signed by 77 Fairfielders on February 7 wielded its Republicanism relatively lightly and did not mention slavery directly, by now Jefferson County Republicans were feeling their strength. On February 15, 1861, Congress acknowledged

---

[125] *Burlington Daily Hawk-Eye*, March 2, 1861, p. 2, col. 3.

the receipt from Iowa's Congressman Samuel R. Curtis of a petition of 400 citizens of Jefferson County, Iowa "praying for the settlement of our difficulties on the basis of the Republican platform." They held that the United States was and should be considered free by default, with slavery limited to the slave states, and not slave by default, with freedom limited to the free states. It was ordered that the petition be "referred to the select committee of five, of which Mr. Howard is chairman."[126]

April 1861: Fairfielders Rally to the Union

When Confederate artillery bombarded United States troops in Fort Sumter in the harbor of Charleston, South Carolina on April 12, 1861, the Civil War officially began. The Southern-sympathizing President Buchanan had downsized the military, and the fort was still unfinished with less than half its full complement of cannon. On April 13, it fell. On April 14, President Lincoln called for 75,000 militia, and on April 16, Iowa was asked to provide one regiment of volunteers for immediate military service.

On April 17, news of the attack and Lincoln's call reached Iowa, and Governor Samuel J. Kirkwood called on Iowa's counties to form militia companies of volunteers to enter military service and help the Federal Government in "enforcing its laws and suppressing rebellion." Each company would comprise at least 78 men, electing one captain and two lieutenants, with ten companies to form a regiment. That evening the loyal citizens of Jefferson County met to martial music, and chose Daniel P. Stubbs as president, Dr. S. W. Taylor and Ward Lamson as vice presidents, and W. W. Junkin as secretary. Christian W. Slagle, J. G. Kirkpatrick, Robert C. Brown, and George Strong gave patriotic speeches, and George Strong, C. W. Slagle, and Robert F. Ratcliff supervised the enrollment of volunteer soldiers.

George Strong enlisted first, followed in order by Moses A. McCoid, David B. Wilson, Henry A. Millen, Robert Lock, George Balding, W. T. Killough, and J. G. Kirkpatrick. A total of 102 men volunteered that evening and over the next few days, and they would form Company E of the 2nd Iowa Infantry, the first in Iowa to take the field.

---

[126] *Journal of the House of Representatives of the United States: Being the 2nd Session of the 36th Congress; Begun and Held at the City of Washington, Dec. 3, 1860...* Washington, D. C.: Govt. Ptg. Office, 1860 [*sic*], Feb. 15, 1861, p. 320.

Through the four years of the war, Jefferson County would contribute over 1,000 volunteers, more than a full regiment. George Strong, first to enlist, would also be the first to die in service, followed by over 200 more.[127]

When the district court began its spring term on April 23, 1861 Democrat George Acheson proposed that the lawyers and court offices renew their oath of allegiance to the government. James F. Wilson, C. W. Slagle, D. P. Stubbs, and Democrat I. D. Jones supported Acheson's proposal. The Democrat Charles Negus did not. He refused to take the oath, and "was subjected then and later to vigorous and often bitter criticism."[128]

1862: 108 Fairfielders Ask Congress to Abolish Slavery

On January 20, 1862, about nine months into the Civil War, the great civil-rights Congressman James F. Wilson—succeeding Samuel R. Curtis, who had resigned to lead the 2nd Iowa Infantry—laid upon

---

[127] Those first enlisting were George Strong, Moses A. McCoid, David B. Wilson, Henry A. Millen, Robert Lock, George Balding, W. T. Killough, and J. G. Kirkpatrick. Bill Hampson, George H. Case, William Scott, Daniel W. Brown, G. H. Myers, A. K. Updegraph, C. A. Miller, W. F. Smith, J. M. Hughes, R. M. Rhamey, Daniel Smith, David P. Long, George W. Hill, John Swanson, Isaac Olds, George W. Fetter, John T. McCullough, D. B. Johnson, John Locke, Manford Hall, Thomas Hoffman, John R. McElderry, Charles J. Reed, N. Howard Ward, David Jones, William H. Cusick, Jacob Fox, J. A. Whitley, W. C. Henderson, Owen Bromley, Samuel B. Woods, William Hill, Brainerd Kerr, James F. Crawford, John J. Payton, R. P. Moore, Jacob Young, Harry Patrick, W. S. Moore, William Leith, H. G. Ross, Matt Hilbert, W. T. Hendricks, McDonald Parshall, Sol. D. B. Welch, William H. Baker, James W. Workman, James Ross, David Pierson, Samuel Turner, George Heaton, William W. Maxwell, John T. Russell, A. R. Wilson, James M. Dudley, Reuben Coop, John J. McKee, Wesley Summers, Silas Pearson, Samuel H. Simms, J. W. Robinson, Elijah Newby, Benjamin Mikesell, Ostin Sebrin, D. W. Garber, Lester Daley, R. G. Forgrave, Wiley S. Simms, John C. Duncan, Daniel Moore, Stephen D. Gorsuch, Jackson Hefner, Henry T. Harris, William Pattison, U. M. Davis, J. W. Messick, W. Bauder, Frederick F. Metzler, J. L Thompson, M. Page, A. P. Heaton, William F. Lowery, Mark F. Carter, Timothy W. Austin, Robert Stam, G. W. Hammond, J. S. Longary, L. D. Boone, W. H. Pierson, Marion York, J. H. Forgrave, James Young, R. B. Partridge, and La Torry Webster. See Charles J. Fulton, op. cit, vol. 1, pp. 327-328. For a roster of Jefferson Co. soldiers in the Civil War, see *The History of Jefferson County, Iowa, Containing a History of the County, its Cities, Towns, &c....*, Chicago: Western Historical Co., 1879, pp. 444-464.

[128] Charles J. Fulton, op. cit., vol. 1, pp. 339-340.

the clerk's table of the U. S. House of Representatives a petition from citizens of his hometown of Fairfield, Iowa "praying for the abolition of slavery, compensating loyal masters." And on January 21 and again on February 5, 1862, Mr. Wilson laid upon the clerk's table a petition by George Acheson and other citizens of Jefferson County—108 in all—asking Congress to abolish slavery in the District of Columbia, and also at least in the rebellious states:

> To the Congress of the United States. The undersigned loyal citizens of Jefferson County Iowa, believing that slavery has been the cause of the present rebellion, and is now its main support; that its removal would rapidly hasten the success of our arms; that you in the exercise of your ordinary legislative functions have the power to abolish it in the District of Columbia, and that as a war measure it may be further abolished at least in the rebellious states, do hereby respectfully ask that you enact a law abolishing slavery in said District, and that you exert ev[e]ry power within your control towards the emancipation of the slaves in the rebellious states, with such provisions, as to time manner, and compensation to loyal citizens or exceptions in their favor, as will be the most salutary and effectual.[129]

On March 13, 1862, Congress passed the Act Prohibiting the Return of Slaves. On April 16, President Abraham Lincoln signed The District of Columbia Compensated Emancipation Act, ending slavery in Washington, D.C. by paying slave owners for releasing their slaves. In the summer of 1862, Congress passed the Second Confiscation Act, allowing Confederates 60 days to surrender or have their land and slaves confiscated. And on January 1, 1863, President Lincoln signed the Emancipation Proclamation, freeing all of the slaves in the rebel states.

---

[129] *Journal of the House of Representatives of the United States: Being the 2nd Session of the 37th Congress; Begun and Held at the City of Washington, Dec. 2, 1861...* Washington, D. C.: Govt. Ptg. Office, 1862, Jan. 21, 1862, pp. 201, 212, 267. Geo. Acheson et al. to the Congress of the United States, [Feb. 1862], 37A-G7.1, House Judiciary Committee, Petitions & Memorials, ser. 467, 37th Congress, RG 233 [D-13]. Endorsement. Cited in Ira Berlin et al., *Freedom: A Documentary History of Emancipation, 1861-1867.* Ser. 1, vol. 1: The Destruction of Slavery, NY: Cambridge Univ. Press, 1985, pp. 176-177.

1861-1862: Last Known UGRR Activity in Jefferson County

As Blacks continued to escape the South even during the Civil War, sometimes with help from the Union Army, UGRR activity in Fairfield probably continued at least until Congress passed the Act Prohibiting Return of Slaves in March 1862, or until the Emancipation Act of January 1, 1863. Possibly the UGRR continued even until the end of the war, as Fairfield still had her "Copperheads" upholding Southern slavery until the South's defeat in 1865.

The last known freedom-seeker in Jefferson County was Robert Winn, who on learning he was legally free halted his flight at the Skunk River near Clay, and went on to fight in Iowa's 60th Colored Infantry for the last two years of the Civil War, later returning to Fairfield to live with his family. Almost every adult Black man in Iowa fought in the Union Army during the Civil War.[130]

Other Possible UGRR operators in Jefferson County

A great many more are good candidates for UGRR activity in Jefferson County because of their anti-slavery principles and UGRR connections. These include Rev. Joseph C. Cooper and Daniel Mendenhall's family, already listed as UGRR operators by the Iowa Freedom Trail Project, but as yet without documentation; Rachel (Coppock) Pierce's brother John Coppock, the strong abolitionist who founded the mill-settlement of Coppock on the Skunk River in the northeast corner of Jefferson County about 10 miles south of Washington; other anti-slavery millers like William Read and Silas Deeds, who like UGRR operator Samuel B. McKain may have helped fugitives across their river; the Reverends Julius A. Reed, William A. Thompson, Charles H. Gates, Levin B. Dennis, and Ashbel S. Wells; Robert D. Creamer, Robert F. Ratcliff, Dr. Jeremiah S. Waugh, Mehitable (Owen) and Gilbert M. Fox, Samuel A. Robb, George Spencer, David Pickering, William Pickard, John Williams, John Andrews, E. S. Gage, Moses Black, Isaac H. Crumly, Asahel H. Brown, Melchi Scott and perhaps his son Charles C. Scott, Daniel P. Stubbs, Hervey Coffin, and many others.

---

[130] Lincoln's Emancipation Proclamation of January 1, 1863 had declared only slaves in the Confederate states to be free, and thus actually freed no slaves at the time, except perhaps for those already on the UGRR, like Robert Winn.

In truth, any of the approximately 800 anti-slavery heroes and heroines in the companion volume, *Who's Who in the Anti-Slavery and Underground Railroad Networks of Fairfield, Iowa*, could have been a UGRR operator, though probably only a few actually were.

About the Who's Who

The *Who's Who in the Anti-Slavery and Underground Railroad Networks of Fairfield, Iowa* gives brief biographies of the known Free-Soilers, abolitionists, and UGRR operators in Fairfield, Jefferson County, and vicinity, in husbands' alphabetical order. Spaces are ignored, so that *Van Dyke* is listed as *VanDyke*. Names beginning with "Mc" are not filed like "Mac," so that *Martin* appears before *McBeth*. Children listed separately always follow their parents' entry, not alphabetically but in descending order of age. Within a given biography, we *italicize* the names of other anti-slavery persons with their own biographies in this compilation.

Stars before an entry denote the person's known degree of anti-slavery involvement:

> * Free-Soilers, those opposing the spread of slavery;

> * * Known abolitionists, advocates of Black civil rights, and adult Underground Railroad operators' close relatives whose own UGRR participation is unknown; and

> * * * Underground Railroad operators, members of known UGRR families, and freedom-seekers on the UGRR.

As Antebellum political positions were quite nuanced, and as our research is ongoing, the star system is crude, inaccurate, and tentative, but we hope may be useful as a quick guide to our current evidence on these heroes of Jefferson County. An entry without any stars denotes a "person of interest," perhaps because of suspected UGRR ties, but one whose personal stance remains as yet unknown.

While the South apparently initiated the Civil War because the North would not countenance its policy of unlimited expansion of slavery, and in defeat lost all its slaves, not every hero who fought to preserve the Union was originally anti-slavery. This biographical dictionary focuses on earlier anti-slavery networks, and notes Jefferson County's Civil War soldiers and sailors only when able to place them in the context of those networks.

The Biographical Dictionary

* * *Mary (Hemphill) and George Acheson*
* *Elmira (Frazey) and Rev. Richard Baxter Allender*
* *Mary E. (Clark), Adaline H. (Templeton), and Jacob H. Allender*
* *Christena Sophia (--) and Anders Gustaf Anderson*
* * *Edna (Crew) and John Andrews*
* *Mary (Bruff) and Benjamin Crew Andrews*
* *Eliza Jane (Sargent) and Pizarro Cortez Arnold*
* *Sarah (--) and Isaiah Armstrong*
* *Almira (Stever) and Rev. Andrew Axline*
* *Lydia (Mechem) and William E. Balding*
* * * *Adriana, Dr. Milton D., Alvin W., and Caroline L. Baldridge*
* *Jane (Barr) and Caleb Baldwin*
* * *Margaret (Langfitt) and Joseph Ball*
* *Joseph Bardine*
* * * *Hannah Baird (Loomis) and Hiram B. Barnes*
* *Sarah (Case) and Thomas Barnes*
Catherine (Williamson) and William H. Barnes
* *Polly (Wheeler) and Roswell Beach*
* *Rachel (Fish) and Obil Orin Beach*
* *Mary (Alexander) and James Beatty*
* *Mary Jane (Kelly) and John Hamilton Beatty*
* *Nancy Catherine (Moorman) and John Bell*
* *Elizabeth Hayes (Ashby) and Hon. William Smith Bickford*
* *Sumner M. Bickford*
* *Mary Jane (Holliday) and Samuel E. Bigelow*
* *Prudence (Strong) and Elijah Billingsley*
* *Temperance (Hood) and Nathan Birkhimer*
* * *Nancy Crockett (Glasgow) and Moses Black*
* *Rachel (Stever) and William Glasgow Black*
* *Margaret A. (Harris) and Joseph Blakely*
* * *Amy D. (Rhodes) and Lewis F. Boerstler*
* *Lucy A. (Brown) and Dr. Andrew Hasket Bronson*
* *Elizabeth (Gray) and Tinley Messick Brooks*

* * *Eliza (Goodenough) and Rev. Joseph G. Brooks*
* * *William Sanford Brooks*
* * *Isabella (Steele) and Asahel Harrington Brown*
* *Sarah Ellen (Fee) and Isaac Harrington Brown*
* *Rebecca (Fletcher) and David C. Brown*
* *Mary Ann (Rhea) and Silas Randolph Burgess*
* *Elcy A. (Smith) and Stephen Butler*
* *Mary (Miller) and Thomas Byers*
* * * *Isabella Bonner (Woods) and Jesse Hoover Byrkit*
* * * *John Byrkit*
* * * *Hannah J. Byrkit*
* * * *Rev. George Washington Byrkit*
* * * *Archibald Reed Byrkit*
* * * *Rev. William Asbury Byrkit*
* * * *Francis Marion Byrkit*
* * * *Martin Luther Byrkit*
* * * *Christian Shaffer Byrkit*
* *Sarah Jane (Young) and George Washington Calfee*
* * * *Esther (Saunders) and Isaac Newton Calhoun*
* *Nancy Parley (Truman) and William Martin Campbell*
* *Mary Elizabeth (Fleenor) and Wesley Bigelow Campbell*
* *Mariah (Campbell) and Nelson Harrison Campbell*
* *Elizabeth (Boyers) and William F. Campbell*
* *Esther (--) and Rev. Borter Hussey Canaday*
* *Esther (Osborn) and Kirby Caviness*
* *Martha (Bonnifield) and Philander Chandler*
* *Missouri (Greenland) and Reuben Chilcott*
* *Wilmirth Jane ("Jenny") (Cox) and Hon. George Miles Chilcott*
* *Adeline A. (Clark) and Solomon B. Clapp*
* *Elisabeth (Albright) and William Dickey Clapp*
* *Jane (Van Nostrand) and Loren Clark*
* *Sarah (McCauley) and Matthew Clark*
* * *Mary Ella (Patrick) and Wickliffe Martin Clark*
* *Sarah Louisa (Wadsworth) and Dr. Charles Shipman Clarke*
* *Mary E. (Gregg) and John Cochran*
* *Elizabeth Jane (Lynch) and George Cochran*
* *Elenor J. (Abrams) and James W. Cochran*
Elmira (Townsend) and Albert M. Coffin
* * *Rachel (Mills) and Hervey Coffin*
* * *Harriet Safronica (Hitchcock) and James W. Cole*

* *Anna Barbara (Johnston) and Nathan R. Cole*
* *Rhoda A. (--) and Charles W. Coleman*
* * * *Margaret (Livingstone) and William Collier*
* *C. C. Collins*
 * * *Jane (Chilcott) and Waltus Collins*
* *Hellen (Powell) and Jefferson Cook*
* *Mary Ellen (Howard) and William S. Cooke (or Cook)*
* * * *Rachel Alexander (Van Dyke) and Rev. Joseph Calvin Cooper*
* *Mary Ann (Patent) and William Harrison ("Harry") Copeland*
* * *Hannah (Neal) and John Coppock*
* * *Lucinda (Crawford) and Rev. Dr. Thomas Edward Corkhill*
* *Elizabeth Edwards (Endersby) and Dr. William Wallace Cottle*
* * *Agnes I. (Thompson) and James J. Cowan*
* *Elizabeth (Holton) and James Dougherty Crail*
**Anna Charlotte (McCaskey) and Benjamin Franklin Crail*
* *Lucretia (Nimocks) and Milton L. Crail*
* *Nancy Jane (Hardin) and George Craine*
* *Charlotte Ann (Shaffer) and James Fletcher Crawford*
* * *Martha (Pettyjohn) and Robert Dobbins Cramer (or Creamer)*
* *Mary Ann (McKee) and Daniel Creegan*
* * *Rebecca L. (Hackney) and Isaac Hammer Crumly*
* *Mary Martha (Beatty) and William C. Cummings*
* *Lucinda Jane (Cummings) and James A. Cunningham*
* *Sybil S. (Snow), Armanda (Parmenter), and Simeon S. Cushman*
* *Mary (Kirkpatrick) and David Daniels*
* *Jane (--) and Dr. Ephraim Darling*
* *William H. Darling*
Sarah Ann (Hurt) and Charles P. David
* *Ann Maria (Judd) and William T. Day*
* *Lucy Ann (Foreman), Martha A. (Stevens), and Silas Deeds*
* * *Betsy Davis (Holloway) and Rev. Levin Beauchamp Dennis*
* * * *Martha Jane (McKemey) and Daniel DeVecmon*
* *Sarah A. (Simmons) and Dr. William C. Dial*
* *Margaret (Baum) and Rev. Joseph R. Dole*
* *Sarah Anne (Miller) and William Downing*
* *Elizabeth (Dill) and John W. DuBois*
* *Vashti (Willits) and John Ellison Dunham*
* *Sarah (Murphy) and William Dunwoody*
* *Mary Ann (Peebler) and William Fletcher Dustin*
* *Eliza Jane (Craine) and Emery E. Easton*

Hannah (Riggs) and Peter Eckley
* *Sarah (Warner) and John H. Ecroyd*
* *Jane (T.) and Aaron O. Edwards*
* *Elizabeth (Heston) and Isaac Ellis*
* *Sarah (Shockley) and Isaac W. Ellis*
* * * *Esther C. (Talbot) and Zachariah Ellyson*
* *Margaret (--) and John Martin Ely*
* *Margaret (Frakes) and Archelaus Moorman Emry*
* *Ruth (Hodson) and Thomas Frazier Emry*
* *Susanna (Hoover) and Samuel Eshelman (or Eshleman)*
* *Mary (Moore) and David Jones Evans*
* *Jane Barron (Ross) and Thomas Davis Evans*
* *Mary Ann (Gregg) and Joseph P. Fansher*
* *Mary Jane (Crawford) and Samuel Carter Farmer*
* *Sarah (Sills) and Joseph Fell*
* *Nancy (Hughes) and John C. Fetter*
* *Sarah Jane (Cloke), Elizabeth C. (Parnell) and George Fisher*
* *Sarah D. (Downey) and John D. Fleenor*
* *Catherine A. (Dole) and Thomas Bracken Fleenor (or Fleanor)*
* *Nancy (Dennis) and Morgan Flower*
* * * *Harriet Jane (Alexander) and Seth Fordyce*
* *Gilbert M. Fox*
* *Margaret Armstrong (Pancoast) and Dr. Benjamin F. Freeman*
* * *Robert French*
* *Mary (Shepherd) and Henry Frush*
* * * *Mary Jane (Armstrong) and George W. Frush*
* *Felicia H. (Lamb) and George C. Fry*
* *Eliza (Jones) and Alexander Fulton*
* *Safronia Augusta (Connable) and Alexander Robert Fulton*
* *Electa (Wallace) and Ebenezer Sumner Gage*
* * * *Jane S. (Dillingham) and Richard Gaines*
* *Elizabeth (Pendlum), Eliza (Anderson), and Rev. Isaac S. Galliher*
* *Mahala (Shaffer) and John Gantz*
* *Mary Magdalena (Bulger) and Isaac Garmoe*
* *Hannah Beraman (Sperry) and Jacob Garver*
* * *Mary (Hobbs) and Rev. Charles Henry Gates*
* * *Sarah (Burton) and Rev. Reuben Gaylord*
* *Nancy (Hart), Nancy A. (--) Brewer, and Ezekiel Johnson Gillham*
* *Mary Sarah (McCulloch) and Thomas Wilson Gobble*
* *Sarah (Stever) and Henderson W. Gorsuch*

* Virginia (Belot) and Stephen Decatur Gorsuch
* * * Mary (Livingston) and Samuel Gould
* * * Rebecca (Brownfield) and Ebenezer S. Gould
* Jennet (Parker) and Peter F. Gow
* Cynthia Anna (Crail) and John M. Grafton
* Eliza (McConnaughey) and James Graham
* Sophia (Flower) and Goodman Graves
* Mary (Smith) and Coleman Graves
* Juliann (Allender) and Archer Green
* Hannah (Taylor) and Marmaduke Green
* Plantena (Allender) and Dr. Wesley Johnson Green
* Harriet Jane (Robb) and Henry Gregg
* Mary Isabel (Parker) and Thomas Griffin
* * * Mary (--) and Peter Groesbeck or Grousbeck
* Winnie L. (Warden) and John L. Hadley
* Christopher Columbus Hall
* Nancy G. (Odell) and Dr. Richard Rozzell Hall
* Sarah A. (Donnally), Margaret A. (Beatty) and Evan C. Hampson
* Sarah (Brown) and George Hanawalt
* Sarah (Killough) and Swain Hand
* Sarah (Jenks) and William Harper
* Minerva (Johnson) and James Harvey
* Elizabeth A. (Hargrave) and Hon. Benjamin A. Haycock
* Elizabeth (Tullis) and Alfred Parker Heaton
* Jane (Taylor) and Adley Hemphill
* Leodica (Baird) and Hiram Milton Henderson
* Nancy (Thompson) and Samuel Henderson
* Elizabeth Kepford (Alt) and James Howell Hendricks
* Susannah (Wood) and Elliott Hiatt
* Catharine (--) and John H. Hill
* * * Julia (Hadley) and Jesse Hinshaw
* * Sarah (Swan) and David Hitchcock
* * Louisa (Rich) and William Parmelee Hitchcock
* * * Caroline (Grossman) and Rev. George Beckwith Hitchcock
* * Charles Rollins Hitchcock
* * Lucy Moulton (Hubbard) and Jared Beecher Hitchcock
* * Caroline Amelia Hitchcock
*Alice Mariah (Schooley) and William M. Hoagland
* Caroline Charity (Webb) and John Hoaglin, Jr.
* Lydia Ann (Moorman) and Rev. Dr. William Zeno Hobson

* *Emeline (Hadley), Julia (Hadley), and Dr. George Arnott Hobson*
* *Amanda Jane (Randle) and Thomas Hodson*
* *Elizabeth (Talbert) and James Frazier Hodson*
* * * *Margaret (Long) and James William Holbert*
* *Mary Ann (Peebler) and Joseph Holsinger*
* *Eliza (Thornberry) and Eli Hoopes*
* *Jane (Nicholson) and John Hopkirk*
* *Phoebe (Walton) and John Bowen Horn*
* *Rev. John Horton*
* *Catherine (Mount) and George Howell, Jr.*
* * * *Herbert Mallory ("Hub") Hoxie*
* *Elizabeth Holmes (Shaffer) and Dr. John T. Huey*
* *Mary Alice (Darling) and Abijah Hughes*
* *Elizabeth (Warwick) and Lawson B. Hughes*
* *Margaret (Johnston) and Robert Smiley Hughes*
* *Louisa (Gorsuch) and Fielding Thomas Humphrey*
* * *Harriet E. (Plumb) and Rev. Robert Hunter*
* *Sarah (Rudy) and Franklin B. Huntzinger*
* *Mary Ann (Smith) and John H. Huston*
* * *Rachel (Tapscott) and Joel Gibbs Hutchin*
* *Victoria (Dutton) and Isaac Hutchin*
* *Elvina Hanna (Russell) and Henry William Hutchin*
* *Margaret Ann (Hanson) and Dr. Nelson Reed Immell*
* *Elizabeth Brown (Brooks) and John Alexander Ireland*
* *Margaret L. (Rea) and Dr. William Penn Irland*
* * *Sarah Louise (Boerstler) and Benjamin Franklin Ives*
* *Sarah B. (Meek) and James Jeffers (or Jeffries)*
* *Mary J. (Jeffrey) and George Washington Jenkins*
* *Annis Hanney (Wilson) and Dr. John J. Jones*
* *William H., Albert R., Harrison, and Arthur Scott Jordan*
* *Sarah (Rambo) and Joseph Junkin, Sr.*
* *Mary Margaret (Cotton) and Joseph Junkin, Jr.*
* * *Elizabeth (Patrick) and William Wallace Junkin*
* *Thomas Lewis Keck*
* *William Keech*
* * * *Jane Irene (Bales) and Mordecai Hiatt Kellum*
* *Malinda (Westfall), Margaret (Abrams), and James O. Kirkpatrick*
* *Rachel Elizabeth (Taylor) and William M. Kirkpatrick*
* * *Celena (Graham) and Samuel Kirkpatrick*
* *Cynthia (Ball) and Gilbert B. Kirkpatrick*

* *Susanna (Comer) and Henry D. Kness*
* *Frances C. (Brooks) and John W. LaForce*
* * * *Selina B. (Byers) and Jacob Taylor Lamb*
* *Maria A. (Danielson) and Ward N. B. Lamson*
* *Sarah (Murrow) and Samuel Waterman Langdon*
* *Sarah Carlisle (Bell), Hester (--), and William L. Layton*
* *John M. Liggett*
* *Catherine Bevier (Nimmons) and Solomon Light*
* * * *Martha (Orr) and Rev. David A. Lindsay*
* *Catharine (Stever) and William Long*
* *Susanna Ann (Baird) and Gilbert Percy Loomis*
* *Elizabeth (Sissler or Cisler) and John Lynn*
* *Elizabeth S. (Hussey) and William S. Lynch*
* *Ann (Brown) and Hon. Sterling Perry Majors*
* * *Lydia Minerva (Hitchcock) and Gordon T. Mallet*
* *Stephen M. Martin*
* * * *Susan Law McBeth and Catherine Christine McBeth*
* *Martha Jane (Bivens) and William R. McCartney*
* *Lucinda (Betterton) and Joseph Benson McClain*
* *Sarah (Irwin or Irven) and John McCleery (or McCleary)*
* * *Hon. Moses Ayers McCoid*
* *Anne Cooper (Slagle) and Francis Burdette McConnell*
* *Sarah Narcissa (McCreery) and Dr. Cyrus McCracken*
* *David Ralston McCrackin*
* *Sarah (Britt) and John McCullough*
* *Mary (Strong) and James McCullough, Jr.*
* *Susan (--) and James McFee*
* * * *Mary Jane (Roat) and Samuel B. McKain*
* * * *Cynthia Ann (Hemphill) and Joseph Alison McKemey*
* *Mary (Fetter) and William Long McLean*
* *Jane (Templeton) and John McLoney*
* *Mary Ann (Goodall) and Isaiah W. McManaman*
* *Martha E. (Russell) and Finley McMartin*
* * * *Emily (Whitacre) and Alfred Meacham*
* * * *Eliza Jane (Parshall) and Dr. Thomas Scott Mealey*
* *Jemima (Graham) and Dr. Edmund Mechem*
* * * *Susannah (Pierce) and Daniel Mendenhall*
* *Emily (Arrington) and Samuel H. Merritt*
* *Rebecca (Haywood) and James W. Messick*
* *Elizabeth (McCormick) and Jacob Metz*

Harriet (Cheadle) and William Miller
George D. Milligan
* * * *Ursula F. (Stone) and Manning Bidwell Mills*
Elizabeth (Parker) and Ralph Richardson Mills
* * * *Almira (Swift) and Hon. Thomas Mitchell*
* * *Maria Elizabeth (Tool) and Hon. Henry Blake Mitchell*
Margaret (Brooks) and Dr. John Tucker Moberly
Amanda Melvina (Frush) and Johnston Moore
Anna (Nimmons) and William Spencer Moore
Margaret (Canaday) and Hon. Thomas Moorman
Parlee (Neal) and Philo Morehouse
* * * *Deanna Adeline ("Dicy") (Barnett) and Henry Morgan*
* * *Matilda (Hollingsworth) and Obediah Morgan*
Phebe (Conkling) and Samuel Mount, Sr.
Maria (Koontz) and John Mount
Elizabeth Jane (Crawford) and Joseph S. Mount
Frances (Bell), Mary Ann (Ramsay), and Dr. Jacob Lewis Myers
Elizabeth Jerusa (Clark) and John Vinson Myers
Dorcas Margaret (Myers) and James W. Nicholson
Amy (Thomas) and Ruel Nimocks
Zilpha K. (Rees) and Christian Emigh Noble
Susannah (Hodson) and Solomon Nordyke
Harriet Eliza (Gilbert) and Stephen Monroe Northup
Mary (Snyder) and Christian Ohmart
Hannah (White) and Thomas Elwood Osborn (or Osborne)
George Whitfield Pancoast
Jane (Johnston) and Col. John Park
Nancy Jane (McCullough) and Lorenzo Dow Parker
Martha A. (Craycraft) and Wilson S. Parker
Caroline Tapscott (Hutchin) and Peter S. Patton
William Patrick
Susannah (Roberts) and David Peters
Mary (Garrett) and John Pheasant
* * *Mary (Bell) and William Pickard*
* * *Agnes (Heston) and David Pickering*
* * * *Rachel (Coppock) and Benjamin D. Pierce*
Rhoda (Burgess) and Robert C. Pleugh (Plue, Plew, Plugh, &c.)
Elizabeth (Austin) and Theron Plumb
Thomas L. Pollard
* * * *James P. C. Poulton*

* Nancy (Derby), Sarah E. (Wright), and Serene Cromwell Pumphrey
* Sarah (Longerbone) and Jacob Ramey
* Elizabeth E. (McComb or McCombs) and Mungo Ramsay
* Elizabeth M. (Bryan) and John P. Ramsay
* * Martha Helen (Pike) and Robert Fleming Ratcliff
* Matilda (Bottom) and Hon. William Moses Read (or Reed)
* Anna Clark (Canfield) and Dr. Charles Reed
* * Caroline (Blood) and Rev. Julius Alexander Reed
* Ann Emelia (Botts) and Ira Graves Rhodes
* * Rev. George Gaby Rice
* Mary Magdalene (Saddler) and John Rider
Zelina (Newell) and Barnet Ristine
* Rebecca (Cassiday) and John F. Robb
* * Mary L. (Hitch) and Samuel Andrew Robb
* Mary C. (Brown) and Thomas A. Robb
* Nancy (Webb) and Benjamin Robinson
* Margaret (Gregg) and Rev. George W. Robinson
* Ellen A. (--) and Christopher T. Robinson
* Leah Beauchamp (Brooks) and Dr. Sawyer Robinson
* Louisa (Pinnecke) and David Rock
* Margaret (Steutzer), Salome (Schoenberger) and Hon. Louis Roeder (or Reeder)
* Mary Jane (McCoid) and James M. Runnells
* Rebecca Ann (Scott) and Artemus Rush
* Sarah (McGowen) and William E. Sargent
* Julia Ann (Boyers) and Abram McLean Scott
* Margaret B. (--) and Entellus McLean Boyer Scott
* * Elizabeth (Clouse) and Melchi Scott
* Mary Eleanor (Thompson) and John Ruth Shaffer
* Melvina Jane (Curry) and Dr. Joshua Monroe Shaffer
* Rachel (Fell) and Orville Olmstead Sheldon
* Joanna Catherine (Woods) and William Emory Sherfey
* Mary Ellen (Uttz) and William Hull Sheward
* Harriet C. (Hammond) and Nathaniel Simmons
* Sarah (Weaver) and William Leroy S. Simmons
* Mary Ann (Howe) and George Washington Sisson
* Amanda (Craine) and James Marshall Slagle
* * Nancy Maria (Seward) and Christian Wolff Slagle
* Susannah Byerly (Immell) and Peter Slimmer
* Mary E. (Woodward) and Alfred H. Smith

*Catharine Beaver (Cowan) and Dr. Gustavus A. Smith
* * * Sarah (Riddle) and Dr. John Jackman Smith, Jr.
* Eleanor (Crail) and John Calvin Snodgrass
* Sarah (Price) and Jacob Snook
* Sarah Ann (--) and Augustus R. Sparks
* * Mary (Gillett) and George F. Spencer
* Angeline (Nace) and Joseph Joel Sperry
* Marie Salome ("Sarah") (Bertsch) and John Spielman
* John A. Spielman
* Nancy R. (Colman) and Hon. Francis Springer
* * * Annis (Haworth) and Allen Stalker
* * * Almedia (Williams) and Charles Osborn Stanton
* Mary Ellen (Dodson) and Daniel Stephenson
* Mary Ann (Wilkins) and Robert Stephenson
* Hannah Amanda (Ball) and Solomon Fink Stever
* Juliet Hastings (Wells) and George Stever
* Mary R. (--) and George W. Stewart
* * Anna (Macy) and Samuel Street
* George Strong
* Mary Melissa (Barker) and James Madison Strong
* * Caroline (Hollingsworth) and Hon. Daniel Parham Stubbs
* Nancy (Stanton), Juliann (Ellis) and Thomas Talbert
* Amanda (--) and Levi P. Taylor
* Amy (Makepeace) and Dr. Samuel W. Taylor
* Lucinda (Summers) and George W. Teachenor
* * Sarah Jane (Thompson) Temple
* * Harriet Boynton (Sawyer) and Rev. William Austin Thompson
* Lydia A. (Shaffer) and Samuel Stark Tipton
* * Alvin Turner
* Eliza (Black) and Benjamin Bishop Tuttle
* Elizabeth Ann (Murdock) and George Washington Vance
* Mary Ellen (Haywood) and Richard Henry Van Doren
* Elizabeth (Giltner) and William Richard Vaught
* Christeann (Schuyler), Mary (Brewer) Drake and Dr. Peter Walker
* Phoebe (Robinson) and William Harrison Walker
* Luzanne or Luzena (Brazelton) and Col. William H. Wallace
Deborah (Thomas) and Josiah Walton
* James Wamsley
* Hannah M. (Woodward) and Jesse F. Warner

* Hannah L. (Page) and Henry P. Warren
* * Deborah (Murray) and Dr. Jeremiah Sturgeon Waugh
* * * Jane (Hadley), Rachel (Mills) and Jesse Britton Way
* Mary (Ream) and John H. Webb
* Martha Louise (Mathews) and Alvin Thayer Wells
* Sophia H. (Hastings) and Rev. Ashbel Shipley Wells
* Caroline (Jordan) and John Henry Wells
* Priscilla Darlington (Holmes) and George Ashbel Wells
* Emily Dyer (Jordan) and Captain William Reed Wells
* Jane (Thompson), Elizabeth C. (Dorland), and Elisha H. Wetmore
* Catherine C. (Mount), Emily E. (Munhall), and James M. Whitham
* * Lucinda Sophia (Butler) and Rev. Reed Wilkinson
* * Harriet (Smith), Martha Ann (Mills) and John Williams
* * Susan (Marsden) and Hon. John Williamson
* Elenor Blaine (Walker) and Grinder Wilson
* Nancy (Briggs) and Robert Wilson
* * Mary Ann King (Jewett) and Senator James Falconer Wilson
* Moses Wilson
* * Elizabeth (Eckert) and William Duane Wilson
* * * Mary Jane (Greenup) and Robert Winn
* Mary (Moss) and William Powell Winner
* Charlotte (Noggle) and John Winsell
* Katherine (Frederick) and Charles W. Wood
* Rebecca (Talbert) and Paul Wood
* Minerva Jane (Smith) and Archibald Reed Woods
* * Mehitable (Owen) Cooper Fox Ellis Woods
* Eliza Ann (Wolph) and John M. Woods
* Mary (Wolph) and Dr. Peter Nesbit Woods
* Samuel Bowen Woods
* Permelia (Hall) and Joshua H. Wright
* * * Nancy (Hale) and James H. Yancey
* Henrietta (Gardner), Clarissa A. (Bloss) and Daniel Young
* Elizabeth (Fell) and Andrew Yount

www.ingramcontent.com/pod-product-compliance
Lightning Source LLC
LaVergne TN
LVHW051842080426
835512LV00018B/3018